A Passion for the Planets

Frontispiece: Tycho Brahe looks skyward at the conjunction of Jupiter and Saturn in August 1563; an event that taught him that the astronomical tables based on the planetary tables of both Ptolemy and Copernicus were unacceptably in error, and led him to devote his energies to obtaining more and accurate observations. This painting is somewhat anachronistic, since it shows the middle-aged Tycho depicted in the statue of him at Hven. In 1563, he was only sixteen. Painting by Julian Baum, © Julian Baum.

William Sheehan

A Passion for the Planets

Envisioning Other Worlds, From the Pleistocene to the Age of the Telescope

Springer

William Sheehan
Willmar, MN
USA
sheehan41@charter.net

ISBN 978-1-4419-5970-6 e-ISBN 978-1-4419-5971-3
DOI 10.1007/978-1-4419-5971-3
Springer New York Dordrecht Heidelberg London

Library of Congress Control Number: 2010927416

Printed on acid-free paper

Springer is part of Springer Science+Business Media (www.springer.com)

In memory of Robert Burnham, Jr., author of the Celestial Handbook series, who quietly stoked it.

Preface

There are many books about the planets, so it is reasonable to ask: Why another one?

With the advent of the first space probes to Venus and Mars almost half a century ago, there has been an explosion of our knowledge about the planets, and it is no longer easy to remember the days when they were blurred and tantalizing images scrutinized at the eyepiece of the telescope. A great deal is now known about the formation and evolution of the planets, their geology and meteorology. Much of this is relevant to a proper understanding of the Earth, which increasingly appears in context as a unique but fragile world subject to cataclysms from outer space and volcanic eruptions, to periodic swings of climate and episodic mass extinctions. These events undoubtedly have deep implications for human survival. Ultimately, they can be understood only in the broader cosmic context of modern planetary astronomy – thus our very survival may well depend on what we have learned about the Earth in relation to its sister planets.

Since the first members of our species appeared in East Africa somewhat more than a hundred thousand years ago, our ancestors have had only a limited knowledge of this cosmic context. For most of that long history, the planets have loomed as distant and inscrutable backdrops to human affairs – watching, without commenting on, the advance and retreat of glaciers and the rise and fall of empires. Yet those wandering lights in the sky have always been objects of beauty and wonder, and especially since the invention of the telescope in Holland and its brilliant employment by Galileo four hundred years ago, they have been the objects, at least among a select few, of single-minded and passionate pursuit. It is the motivation and passion the planets have inspired in at least some of us to study them that is the main and unique subject of this book. It is something – like the passion for chess – that is not, presumably, shared by any other species on this planet; but while this intellectual preoccupation is uniquely human, the passion for the planets no doubt shares the underlying neural machinery of ancient drives – tropisms for light and nutrients – that caused even single-celled animals to point themselves upwards toward the sky and the saurian ancestors of birds to take to the air.

Like other similar drives that are sought for their own sake rather than for secondary gains or rewards, the study of the planets has never been lucrative. Some of those who studied them – like Tycho Brahe and Percival Lowell – have been blessed with wealth and means allowing them to pursue their idiosyncrasies in comfort and with the best instruments. Many more have struggled, sustained only by the intrinsic rewards of the pursuit itself, and willing to make great sacrifices in its behalf. They have done so because they could not help themselves, and that, in some ways, is what defines theirs as a great passion.

I know of nothing more satisfying than to stand beneath the stars, with a brilliant planet shining in the sky. Then all worldly cares fall away, and one lives – for a moment – in the shadow of eternity. There is an exhilaration to communing with other worlds that is addictive. Moreover, for those who follow the movements of the planets, the rhythms and seasons of life have always been played out against elongations and oppositions, conjunctions and occultations. Such events are transcendent, and give a sort of majesty to existences that otherwise can seem so trivial and ephemeral. I will have succeeded in my aims if I have given some hint of the romance that the lover of the planets feels – what prompts his or her continued interest – rather than only what he or she knows.

I have been helped by many people to write this book. I owe a debt to those who – after I first began to wonder at the moving lights in the sky as a boy of nine – nourished, shared, and sustained that interest.

Robert Burnham, Jr., was among the first; he wrote to encourage me at a time when I thought there could be no grander destiny than to be employed, as he was, as a researcher at Lowell Observatory – a place I already loved because of its associations with Percival Lowell and life on Mars. Receiving his letter was one of the great thrills of my life. But childhood is a dreamy time, indifferent to practical realities, and I later learned that Burnham's situation was far different than I imagined. By the time I had matured enough to truly thank him for what he did, he was no longer at the observatory but writing (in *Sky & Telescope*) with a mind unhinged: "I have devoted over two decades of my life to astronomy, and my celestial handbook has been called a modern classic. I am the discoverer of six comets, not to mention thousands of new proper-motion stars, which my colleagues and I found during a 21-year proper motion survey at Lowell Observatory. As a result of all these accomplishments, my income has rarely risen much above the poverty." That was a great shame. Regrettably, I had waited too late to thank him – he wrote that bitter letter in 1982, at the same time I came to Lowell Observatory to begin my researches in the history of astronomy – and though I made some queries, no one knew how I might reach Mr. Burnham by then. So I never did, and though it is far too late, I nevertheless dedicate this book – posthumously – to him.

I acknowledge the contributions of Mike McGarrity, a grade school friend who was the first person with whom I regularly shared my love of the planets – he was a kind of Sidney Ewart to my George Proveno during a number of crucial years – and Mike Conley, who caught the same infectious enthusiasm when we were in high school together, and who has remained active; I observed the planets and stars with him as recently as last weekend. I also express my gratitude to the late Doreen Savage, a high school English teacher who encouraged and greatly cultivated my writing, and to the late Jim LeClaire, M.D. a fellow climber of mountains if only occasionally a scaler of other worlds.

I have been helped and encouraged by many great historians and observers; the following is a partial list: Bill Hoyt, Art Hoag, Mike Crowe, Clyde Tombaugh, Richard Baum, Harold Hill, Richard McKim, Albert van Helden, Andy Young, Dale Cruikshank, Stephen James O'Meara, Patrick Martinez, Henri Camichel, Audouin Dollfus, Don Osterbrock, Owen Gingerich, Rem Stone, Laurie Hatch, Tony Misch, Marilyn Head, Phil Barker, Luigi Prestinenza, Tom Dobbins, John Westfall, Masatsugu Minami, and Dava Sobel. All have contributed in ways too numerous to name but that they will recognize in the pages of this book. I have gleaned valuable insights from many non-astronomers who have provided perspectives on the historical and cultural contexts of the human studies of the planets – especially from Mark Watts, David Lewis-Williams, and R. Dale Guthrie. Julian Baum and Randall Rosenfeld provided exquisite illustrations especially for this book. I also express my appreciation to Melita Wade Thorpe, who organized astronomical tours of Italy and France giving me the opportunity to visit and photograph many important sites and works of art in anticipation of the International Astronomical Year, and the other members of the tour who shared with me their enthusiasm, insights, and sore feet.

Tomasz Mazur kindly provided me his evocative photographs of Copernicus, Frombork Cathedral, and Copernicus's Tower and granted me permission to reproduce them. Richard McKim sent me two photographs of sites in Prague.

I wish to express my sincere appreciation to Harry Blom, at Springer, who has been an ideal editor. He believed in the project from the first, and patiently tolerated my apologies and excuses as it inevitably expanded beyond my initial aspirations and fell tiringly, if never quite hopelessly, behind schedule.

As always, my family deserves the last – and deepest – tribute. My parents, Bernard and Joyce Sheehan, supported my idiosyncratic pursuits with my telescopes with kind encouragements, and never begrudged me my late night and early morning vigils beneath the night sky. My brother, Bernie, often shared those vigils. The memory of my grandparents, Alice and William Robinson, and uncle, William Robinson, Jr., will always be wrapped in the nostalgia of my first telescopic views of the Moon and planets, which I obtained from beneath the noble oaks of the front yard of their modest home. My wife, Deborah Sheehan, has been the companion of my adventures, as well as a mentor to my prose. My sons Brendan and Ryan have eagerly joined me in my astronomical adventures and have their own passion for the planets. Certainly they and their generation will need the broad perspectives of planetary science if they are to salvage this fragile Earth whose humble place in the cosmos astronomers, including some of those described in this book, first defined.

Willmar, MN
October 26, 2009

William Sheehan,

Contents

Chapter 1
Beginnings

I should not talk about myself so much if there were anybody else whom I knew as well.

Henry David Thoreau, *Walden*

I grew up in a religious household, where the only book that was felt to be really necessary was the Bible. I learned that the body was something to be despised and that, like my own personal body, the Earth itself – which had been made something like 6,000 years ago – was tinder to be destroyed in a soon-to-be conflagration.

When I was asked to contemplate a realm of greater perfection than this earthly one, where too often desires were frustrated and wishes unheeded, my thoughts were steadfastly directed upwards – to Heaven. As is typical of the tendency of the child's understanding to conflate like concepts, I did not readily distinguish *Heaven* from *the heavens* – the latter term referring to that realm in which the stars, seemingly imperishable, turned in their courses forever. So the heavens were for me something perfect and worthy of immortal longings.

They were everything this world was not. They presented a lovely phantasmagoria of passing clouds, of sun and moon and stars. They were realms of dream.

For a long time I remained rather confused between the religious teachings that were pressed upon me and these fumbling and groping reaches for a perfect world – or worlds – that existed beyond the Earth. I knew stars, generically; but nothing of stars in particular. I did not know the name of a single one. When I looked up at the night sky, I did not, as far as I remember, recognize specific constellations – nor did anyone then or later bother to teach me any of them. Though I must have seen bright planets, just as I did the Sun and the Moon, I did not know their names.

At this time – as a 6 or 7 year old – I still seem to have believed, as far as I remember or can reconstruct, the literal truth of the words of Genesis (1:1):

> In the beginning God created the heaven and the earth.
> And the earth was without form, and void;
> And darkness was upon the face of the deep…
> And God said, Let there be lights in the firmament of the heaven to divide the day from the night; and let them be for signs, and for seasons, and for days, and years…

W. Sheehan, *A Passion for the Planets: Envisioning Other Worlds, From the Pleistocene to the Age of the Telescope*, DOI 10.1007/978-1-4419-5971-3_1,

The author – aged about eight – as he looked about the time he went splashing home from school and first fully grasped (a true eureka! moment) that the Sun was a star. William Sheehan collection

I suppose I had a typically anthropomorphic view of God at the time; he was an old man with a long white beard.

I wasn't a particularly unhappy child. I certainly did not experience the adversity of an E. E. Barnard, who grew up in grinding poverty in the Southern United States after the Civil War and who would lie out in an old wagon bed at night watching the stars pass overhead, forgetting for awhile a life which, he recalled years later, "was so sad and bitter that even now I cannot look back to it without a shudder."

My maternal grandmother, Alice – in whose front yard I would later set up my first telescope and marvel at the remarkable incongruities of the Moon's surface – was born and grew up on the Mesabi Iron Range in Northern Minnesota. Every summer my family went up there for a long visit.

As I was brought up among devout practicing Catholics (my father was Irish-Catholic, my grandmother Slovenian-Catholic) I always went, without fail or excuse, to Sunday Mass. When I was visiting Northern Minnesota I attended Mass just as I did at home in Minneapolis; there was no change or stinting of the inflexible routine. I remember the bald-pated priest, with his stentorian voice, sermonizing as other priests had in the urban settings with which I was familiar (this priest had converted an uncle to Catholicism). The church boasted the usual stained-glass windows and scenes from the life of Christ and images of the saints. However, in the background were different scenes from those of the Inner-City neighborhood in which I grew up: rust-iron hills where the miners worked (my grandmother's father had been a miner; some of her brothers had also worked in those mines). I had always been taught that the Catholic Church was eternal, because it had been founded by Christ himself. It had endured 2,000 years – all of that! To the mind of a child, 2,000 years is a long time. Adding up all the biblical begots – working out the calculations as Archbishop Ussher of Armagh had once done – one got back to an age of 6,000 years for the age of the Earth itself. A long time, 6,000 years, from the perspective of a 7-year old. This old world was creaking and decrepit indeed!

Yet there were geological signboards posted about which described changes supposed to have taken place in these rust-iron hills over millions and even billions of years (these were pre-Cambrian vistas). They spoke of convulsions and revolutions of the Earth; of seas of rust that had settled out over long ages onto old sea-bottoms and been covered layer by layer in silt and limestone and shale; shouldered upward into mountains, they had then been weighed down and sanded off by the wasting combinations of ice and water and wind until these ancient sediments once more stood exposed at the surface, just as they had two billion years ago. Such were the changes wrought in the vast cycles of time!

Though at that age, I had only a faint grasp of numbers that big – a two with nine zeros after it was something to conjure with, and might as well have been a zillion as far as I was concerned – I nevertheless realized that there was a lot of time that was unaccounted for, and wondered why (like sex) nobody ever talked about it.

It was most unsettling, and I had a hard time getting it out of my mind. Frankly, I was terrified by this Old World. Though I did not think of it much during the day, it came back to me with a vengeance at night, when I lay awake thinking of the old sea-monsters swimming through unfathomable deeps in a world not made for man or for me. I tried to shut these speculations out of my mind (as long as I can remember, I have been an insomniac – a clear advantage, incidentally, for someone destined to take up astronomy!). I remember frantically reciting my prayers, rehearsing (with feigned assurance) the dogmas of the Church, trying desperately to use all the bead-like talismans and repetitive mantras I knew to protect me against what I perceived to be crushing and mortal dangers....

I now realize that at the age of seven – or it may have been eight – I had undergone a precocious but serious crisis of faith.

There were marginalia to the biblical Scriptures, I had discovered, which no one ever talked about: margins that were wider than oceans around a floating leaf. Those marginalia included vistas of vast time discernible from these worn rust-iron hills. Though I did not know any of the details, I could at least vaguely grasp their import. (Strange, but no one I knew seemed to know anything about this. My parents, as well as the religious authorities, seemed somehow able to compartmentalize their minds; they kept these props politely offstage and argued with the disingenuous attitude of the Victorian woman who had learned of the first discovery of Neanderthals: "I hope it isn't true, but that if it is, no one finds out about it.")

They spoke only of Man. The pre-Cambrian times were not mentioned at all; but the testimony of those geological signboards, and those rust-iron hills, could not be disputed. I did not yet have a clear idea about any of this, but I was already dimly – and uncomfortably – aware that the story of Man was a mere postscript to the pre-Cambrian and whatever else had come before.

> What *is* man, that Thou shouldst be mindful of him?

Where was there a safe-harbor or resting-place for a soul like mine – tiny, unhoused, untethered – suddenly cast adrift into vast seas of limitless *time*?

I did regroup, and managed – once settled back into my usual routines – to put out of mind for awhile these dizzying prospects. My preceptors were able to anesthetize my painfully awakened sensibility and reconcile me, however temporarily, to the assurances of religion (though the kind of fire-breathing religion of damnation and eternal punishments in which I was raised did not necessarily provide *much* comfort). But at times my restlessness to know about forbidden things returned.

Then, at the age of nine, another transformation occurred in my consciousness. This time I am sure of the date. The intense emotion I experienced produced a "flash" or "arousal" memory, and it is even now as vivid as if it had happened only yesterday.[1]

On February 17, 1964 – like an autistic savant, I remember not only the date, but the day of the week, a Monday – I was walking home from St. Bridget of Ireland Catholic School. I was in Fourth Grade. We were having a first winter thaw; a delicious early spring-like breeze caressed my cheek, and the snow was melting and ponding into puddles. A child's delight! I blundered artlessly through the stands of water, splashingly, with my boots on, as I had done on similar occasions before. Perhaps, on this genial and pleasure-charged day, I was more aware than usual of the presence of the Sun; of the image of its round perfect disk reflected in these pools of water and the dramatic circles of light breaking, as if by magic, into electric dots and dashes – codes of wonder – as I splashed through them. For whatever reason, something occurred to me in a flash of awareness; I realized, as Giordano Bruno had realized – Bruno, burned on February 17, 1600 as a heretic by what was then my own Church, though I had not yet heard of him – that the Sun is a star, and the stars are distant suns. (Of course, I must have *heard* this; but it had never been real to me, I had never taken it in.) I was bowled over. That perception of reality compelled my imagination. I stood awestruck, and looked up astonished. I had blundered into something truly portentous.

[1]Memory fidelity is intimately related to high-arousal episodes. Obviously, what I experienced on this day was a high-arousal episode. On memory fidelity and high-arousal episodes, see: P.E. Gold, "Sweet memories." *American Scientist*, 75 (1987):151–155.

Statue of Giordano Bruno in the Campo del Fiori in Rome, on the very site where he had been burned at the stake as an unrepentant heretic in February 1600. This statue was erected in 1889 after an international plea for funds and support. Subscribers included Victor Hugo in France, Herbert Spencer in England, Ernst Haeckel in Germany, and Henrik Ibsen in Norway. Photograph by William Sheehan in 2001

Joseph Campbell, in *The Hero with a Thousand Faces*, notes that one of the ways an adventure can begin is through

> a blunder – apparently the merest chance – [which] reveals an unsuspected world, and the individual is drawn into a relationship with forces that are not rightly understood. As Freud has shown, blunders are not the merest chance… The blunder may amount to the opening of a destiny.[2]

[2] Joseph Campbell, *The Hero with a Thousand Faces*. New York: Pantheon Books, 1949; p. 51.

No doubt I had heard – as a matter of mere indifferent knowledge, as a statement of fact – that the Sun was a star. But if so, until this moment it had never been *real* to me. Now I experienced a *eureka* moment. I appropriated the fact, stamped it with stinging reality into myself. William James, in his *Principles of Psychology*, discusses what he calls the perception of reality:

> The *fons et origo* of all reality, whether from the absolute or the practical point of view, is … subjective, is ourselves. As bare logical thinkers, without emotional reaction, we give reality to whatever objects we think of, for they are really phenomena, or objects of passing thought, if nothing more. But, *as thinkers with emotional reaction, we give what seems to us a still higher degree of reality to whatever things we select and emphasize and turn to* WITH A WILL.[3]

The fact that the Sun was a star I had selected and emphasized and turned to with a will. I had apprehended that fact with *passion*, and my life would never be the same again.

For some reason I did not feel any of the intimidation, dread, and sense of foreboding I had felt when I had had those intimations of the vastness of geologic time. To this day I cannot fully account for the difference in my reaction. It may have been no more, perhaps, than the difference in my own mood. During my visits to the rust-iron Precambrian hills I had been homesick and sad and lonely; on this warm February day I was alive in every pore with the promise of an early spring – exhilarated with a sense of freedom such as I had always felt wending homeward from school's grinding routine. In my new enthusiasm I galloped home, and looked up every word related to astronomy I could think of in the Dictionary. I copied each entry down in turn, and eventually I put together a little book.

After writing for a page or two, I looked out the back window – dangerously, foolishly – until my eyes were dazzled at the late-afternoon Sun. I was Icarus flying too close to the Sun until his waxen wings (my tender eyeball) almost melted. I did suffer for sometime afterward from solar retinitis, which included weakening and drying of my eyes as well as headaches, the same sorry condition that all beginners in the ways of astronomical knowledge are warned against in guide-books (and so I feel obliged to say: do as I say not as I did. I also emphatically repeat the warning: NEVER LOOK AT THE SUN WITHOUT PROTECTION FOR THE EYES. Fortunately, I did recover eventually.) Next I looked long and hard at the slender sliver of the Moon settling into the still leafless branches of an elm, now long gone, that then stood in the backyard.

Many astronomers have dated the first dawning of their interest in the skies to the appearance of something extraordinary, such as a dramatic eclipse or a majestic and noble comet. Mine was born through a flash of insight into the nature of the

---

[3]William James, *Principles of Psychology* (New York: Dover, 1950 reprint of 1890 edition); vol. 2, pp. 296–297.

Earth's ruling star on an otherwise ordinary late-winter afternoon. As in my case, so in that of many others who are awakened to passion for the skies, the decisive moment occurs in the prepubescent period, when one is not yet under the sway of the powerful hormonal influences of adolescence, the intense narrowing of consciousness to the all-mastering biological urges of the race. It is then that the passion for adventure stirs and draws one out of oneself, draws one into the Great Beyond with an irresistible siren's call.

When my passion for astronomy was first stirred at age nine – a flame, once kindled, that can rarely be put out – the planet Venus was appearing as the Evening Star. It was slowly working its way toward its Greatest Elongation East of the Sun, which it reached in mid-April 1964.

It became a compelling pastime to find out how early I could first glimpse its sharp pinprick against the blue-grey sky. My eyes, despite the hazard I had subjected them to, were still young and sharp and resilient, and I detected the planet early and discovered I could even make it out in broad daylight. It became my obsession to follow it for hours as it went down, growing more brilliant and lamp-like against the darkening sky until, burying its head near the horizon, it extinguished itself like a burning taper in the vaporous mists and I could follow it no more. I wondered at this lovely planet with its beautiful lucid clear light with an intensity of interest that no Sumer-Akkadian or Meso-American Venus-worshipper could have surpassed! I cannot count the number of times I wished myself far from my troubles in this world into that seemingly perfect, eternal, and untroubled circle of light.

And so I did later with the other planets whose wandering orbs I managed to enmesh and snare in the nets of my curiosity across the open fields of the night sky – royal and resolute Jupiter, golden and stately Saturn, ruddy and temperamental Mars – the last I was most eager to pursue, but it was uncooperatively lying dim and obscure nearly on the far side of the Sun from me.

I became a schoolboy Dr. Faustus, conjuring magic spells beneath my newly discovered starry heavens:

> Now that the gloomy shadow of the earth,
> Longing to view Orions drisling looke,
> Leapes from th'antartike world unto the skie,
> And dimmes the welkin with her pitchy breath:
> Faustus, begin thine incantations…
> Then feare not Faustus, but be resolute,
> And trie the uttermost Magicke can performe.[4]

(Marlowe, *The tragicall Historie of Doctor Faustus*)

[4]Christopher Marlowe, *The tragicall Historie of Doctor Faustus* (1588/89?), in: *The Works of Christopher Marlowe*, ed. C. F. Tucker Brooke. Oxford: Oxford at the Clarendon Press, 1910, lines 235–239 and 248–249.

Like Faustus I soared in imagination, "seated in a chariot burning bright,/Drawne by the strength of yoky dragons neckes." The sense of movement, of exhilaration, as I opened the sail-broad vans of my imagination was as real as any experienced in actual flight.

Ah, what glorious days – and nights – those were, when I became an infatuated lover of these other worlds. On enchanting balmy April nights, with the birds serenading and the wind whispering, I longed for Venus, my serene one, to appear. When it came, the sight of it soothed and tranquilized my soul at the same time it excited and stirred up my ambition. I did not know Wordsworth's poetry at the time but the following lines express something of my sentiments:

> O most ambitious star! An inquest wrought
> Within me when I recognized thy light;
> A moment I was startled at the sight;
> And, while I gazed, there came to me a thought
> That I might step beyond my natural race
> As thou seemst now to do: might one day trace
> Some ground not mine.[5]

The ground not mine was the whole universe beyond the Earth.

I had found an escape from the drudgery of endless rounds of uninspired schoolwork and from the threat of bullies lurking in narrow alley-ways; I could escape, almost at a whim, from the unpredictable moods or capricious rules of many of the adults who surrounded me. I was henceforth set apart – my face shone, though others saw it not. I was a traveler among the planets.

Since I couldn't afford to buy books then, and there weren't any worthwhile ones around the house, I made straight for the nearby public library, located in a romantic old stucco building on a pond near my house. It had been built in 1910, the year of Halley's Comet, with funds provided by the parents of a wealthy youngster named John Deere Webber who had drowned in a creek nearby. It contained a small shelf devoted to astronomy. That shelf soon became my corner of the collection.

The shelf was dominated by the prominent British astronomy author Patrick Moore who, as he told me 40 years later, had himself been "hooked" on astronomy at age six when, on a dull rainy day, his mother put in his hands G. F. Chambers's book *The Story of the Solar System* (1898). I too was hooked and devoured every astronomy-book I could lay my hands on: Moore's *The Story of Astronomy*, H. P. Wilkins and Moore's *How to Make and Use a Telescope*, Robert S. Richardson's *Exploring Mars*, Willy Ley's *Watchers of the Skies*, Tony Simon's *The Search for Planet X*.

I read with keen interest stories now so worn into familiarity that the discovery of any new detail is an event. But they were then encountered for the first time. I read of Ptolemy, who usually came off poorly in the accounts I read with his system of epicycles, created ad hoc to keep the planets centered on the immobile Earth; Copernicus who received on his deathbed a copy of his great book showing that the

[5] William Wordsworth, "It is no spirit who from heaven hath flown"; composed 1803, published 1807.

Sun was the true center of what henceforth was called the Solar System; Tycho Brahe, who lost his nose in a duel and whose passion, stirred by the successful prediction of an eclipse, led on to a career devoted to the stars in which he would don his princely robes before sauntering into the night. Tycho discovered a new blazing star in the W-chair of Cassiopeia and tracked the courses of the planets across the sky with the naked-eye (as I began to do). Mars in particular caught Tycho's eye. He was one of the first *areo-philes* – lovers of Mars – and from his observations his weak-eyed assistant Kepler discovered the true shape of the planets' orbits to be ellipses.

I felt a kinship with all of them, since they were all passionate seekers who lived, as I did, for the planets and stars. I did my best to emulate them. My equipment was necessarily primitive, but though inexpensive, it was excellent. It consisted of my own eyes. Though I could see no farther into the depths of space than Tycho Brahe had done, there was a great deal to see; I was kept busy learning the names of stars and constellations and tracking the planets among those of the Zodiac. Tycho was my first hero.

As beautiful as the planets were to the unaided eye, eventually – and inevitably – I began to long for more. I longed to bring them up close and study them in detail. Galileo, with his small telescope and courage in opposing the tyranny of the Church, soon became another hero of mine. I desperately wanted a telescope of my own. Alas, I could ill afford it. But resolutely I saved my meager allowance (25 cents a week) and worked various odd jobs for my parents until finally I commanded enough money to purchase a department-store telescope – a 2.4-in. refractor I got for the princely sum of 30 dollars.

I shall never forget the first night I used it – September 12, 1964. I set it up among the oaks of my Slovenian grandmother's front yard and pointed it at the 5-day old Moon standing among the stars of Scorpio. Eagerly I put my eye to the peephole of the eyepiece. What I saw could not have been more astounding. The Moon in that telescope was a revelation. It loomed as a battered world that had endured some remote and terrible apocalypse. Gone was the Moon as it had hitherto been, round and small enough to catch within the palm of my hand, and gone was the homely face of the Man in the Moon. All of this vanished, and something wonderful appeared in its place – a New World. I was no longer seeing a small globe from a quarter of a million miles away but was seeing that globe as it might look from only a few thousand miles away, as it might look were I hurtling at it with H. G. Wells's Cavor and Bedford in their Cavorite-covered sphere:

> The whole area was moon, a stupendous scimitar of white dawn with its edge hacked out by notches of darkness, the crescent shore of an ebbing tide of darkness, out of which peaks and pinnacles came climbing into the blaze of the sun.[6]

[6] H. G. Wells, *The First Men in the Moon*, edited with an introduction by David Lake (New York: Oxford University Press, 1995), p. 47.

The author's venerable first telescope – a 2½-in. refractor – with which he observed the craters of the Moon and the satellites of Jupiter. William Sheehan collection

Over the next few months, I continued to follow in Galileo's footsteps with my precious instrument, which was much more powerful than his. I caught the four large planetary-sized moons of Jupiter he first saw at Padua in the winter of 1610, and did as he did in tracing their shifting places night by night. Then I watched Venus on its next return to Greatest Elongation after the one during which I had first observed it. I saw it as a small disk in the telescope and watched it grow larger and pass through phases like a small lovely moon. Those phases vindicated Copernicus's scheme of the Solar System over Ptolemy's geocentric system. I saw it all with my own eyes. It was splendid and holy.

The joy of those early observations can never be equaled again. Though the latest Hubble Space Telescope observations indicate that there may be six billion Jupiters in our Galaxy (and how many Earths!) the sight of none of them will ever give me as much satisfaction as the sight of the Venus or Jupiter of our own Solar System as seen through my small telescope in the winter of 1964–1965. One has that experience but once in a lifetime.

And so, at the tender age of ten, I had become a hopeless addict to the sensations of the night-sky. Novelty and excitement causes dopamine to surge and thrill the brain; endogenous opiates produce a warm rush of pleasure and contentment – the "observer's high" – lulling me at the end of an observing session into pleasant dreams. No doubt Testosterone also played some role in all this. Though at 10 and 11, I was not yet close to the peak of Testosterone production (which occurs between 17 and 19 and usually narrows one's interest to one thing and one thing only – and it is not astronomy), presumably I was already experiencing some of its effects. Testosterone increases the tendency to take risks and the thrill of danger and adventure that – among our ancestors – was tied up with the quest to hunt challenging game (in the Pleistocene, Great Mammals such as bison, wooly mammoths, aurochs, bears). It also enhances concentration – sometimes to the point of obsession. It increases persistence and attention and focus on the things of interest.

In my case I *was* a hunter – pursuing the game with single-mindedness of purpose – but the game consisted not of animals but rather the deathless behemoths of the sky, which I sighted on not with the point of a spear or the bridge of a bow but with the sights of my telescope. Though there was some sense of risk and danger in being alone in the night, especially living as I did in an inner-city neighborhood, and I was sometimes scared at an unfamiliar sound in the night or the rustle of a bush, I confess that most of the time I was completely safe and the risks – and excitements – were cerebral.

The action of powerful chemicals on the brain's reward-center is strongly reinforcing and so, when I could not get at my telescope (as on cloudy nights), I experienced acute withdrawal. Inevitably, I also developed tolerance: over time I became rather inured to the same sensations, needed the greater rush provided by a larger telescope and a more discerning view.

My passion for the planets had begun. At times it would be so all consuming that it narrowed my behavioral repertoire – like the drug pursued by an addict (though fortunately, in this case, one that was without evident toxicity). I was all for the planets! Nothing else mattered.

Chapter 2
By Passion Driven

O Solitude! if I must with thee dwell,
Let it not be among the tumbled heap
Of murky buildings; climb with me the steep,
Nature's observatory....

John Keats, "To Solitude"

The 2¼-in. department store refractor with which I made my first observations of the Moon and planets was the kind serious hobbyists view with disdain. But it was the best I could afford at the time, and it was worth every penny – it brought me countless pleasurable hours as I followed in the tracks of the Master, Galileo.

My budding astronomical interest was then very far from being exclusively or even mainly intellectual. It was mainly emotive, the whole subject being enveloped, as it were, in a gauzy haze of romance. As I watched the mountains and craters of the Moon pass between realms of light and shadow, charted the movements of the satellites of Jupiter, and followed the phases of Venus (which offered "ocular proof" of the most vivid and convincing kind of the truth of the Copernican system), I might well have said with Thoreau:

> Talk of mysteries! Think of our life in nature – daily to be shown matter, to come in contact with it – rocks, trees, wind on our cheeks! The *solid* earth! The *actual* world! The *common sense! Contact! Contact! Who* are we? *where* are we?[1]

After making contact with these other globes in space – seeing them, almost *feeling* them, as they presented the aspects of balls illuminated by the lamp of the Sun – I knew *where* I was. I was on a moving planet, scurrying around the Sun among the other moving planets. *Who* was I? An inhabitant of the Earth – the Solar System – the Galaxy – the Universe. I *knew* this by age ten.

Those were heady, glorious days for me. A shy, nerdy kid – socially awkward – I loved from afar. The objects of my worship were not flesh and blood creatures but radiant planetary divinities. I was often out at night with my telescope in search of a thrill, and nothing thrilled me more than the sight of these serene lovely worlds,

[1] Henry David Thoreau, *The Maine Woods, Ktaadn* (1848).

W. Sheehan, *A Passion for the Planets: Envisioning Other Worlds, From the Pleistocene to the Age of the Telescope*, DOI 10.1007/978-1-4419-5971-3_2,

jewel-drops of pure and dreamy light. I can hardly express how beautiful they seemed to me. I sought mystical experiences – ecstasies – from my backyard as I was transported from thence across trackless voids. There were moments when I became so enthralled in my pursuit of planetary detail that I almost "forgot myself a man." I was like the Inuit shamans whose capacity for ecstasy, according to the great authority on the subject –

> enables them to undertake any journey "in spirit" to any region of the cosmos. They always take the precaution of having themselves bound with ropes, so that they will journey only "in spirit"; otherwise they would be carried into the sky and would vanish for good.[2]

I too remained safely tethered by my corporeal body to the ground while my mind followed the far-flung transport of my eye to the Moon and other planets. My body was often as still as I could make it as I concentrated on – and attempted to hold fast – the tremulous image. Like them, I was never more fully myself than when I was taking part in such exploits:

> The Eskimo shaman feels the need for these ecstatic journeys because it is above all during trance that he becomes truly himself; the mystical experience is necessary to him as a constituent of his true personality.[3]

Until then it may well be that I had never experienced an authentic spiritual experience. Hitherto I had been at most what William James calls the "ordinary religious believer, who follows the conventional observances."[4] Henceforth I began regularly to experience what might be called "mystical states of consciousness," during which I was transported once and for all outside the conventional, domesticated, second-hand trappings of religious orthodoxy in which I had been raised into the vast untamed world of wild and primordial forces that, as I now know, were as old as the Pleistocene. The planets and stars were wild places; they were not yet domesticated, they summoned me back to the magic of a time when

> … holy were the haunted forest boughs,
> Holy the air, the water, and the fire.[5]

The Earth had once been holy. The planets and the stars were so still.

What curious child can resist looking into a keyhole or peephole or knothole to see what lies therein? Who can resist the invitation to exciting – perhaps even forbidden – knowledge? The purveyor of carnival amusements knows too well the answer. The eyepiece of a telescope is, obviously, such a keyhole or peephole or knothole. Wordsworth caught something of the mood that attended such a novelty in Leicester Square, London, in 1806, in his poem "Stargazers":

---

[2]Mircea Eliade, *Shamanism: Archaic Techniques of Ecstasy*. Princeton, NJ: Princeton University Press, 1964, pp. 292–293.

[3]Ibid., p. 293.

[4]William James, *The Varieties of Religious Experience* (New York: Modern Library, 1936), p. 7.

[5]Keats, "Ode to Psyche," lines 38–39.

> What crowd is this? what have we here! we must not pass it by;
> A Telescope upon its frame, and pointed to the sky....
> The Show-man chooses well his place, 'tis Leicester's busy Square;
> And is as happy in his night, for the heavens are blue and fair;
> Calm, though impatient, is the crowd; each stands ready with the fee,
> And envies him that's looking; – what an insight it must be!

Of course, nothing could be less the setting for a trance than this. Inevitably, the gapers returned disappointed.

The true astronomer – like the mystic and the shaman – travels alone. It could not be otherwise. He is the solitary figure in the darkness. He is the hunter who stalks his game silently, patiently, with a single-minded concentration that is often (necessarily) narrowed to the point of obsession.[6] He is one who perhaps – like the Eskimo shaman – needs to be removed from the distractions of the mundane world by a curtain; or he is the adventurer who relishes nothing more than the solitude at the top of a mountain with the whole world before him and the Big Sky above him for company.

To seek the planets and the stars is to seek solitude from the madding human crowd – it is to stand apart and aloof of necessity taking the attitude that one does not love mankind less but nature more.

Almost all astronomers I know are at heart loners. They love heights, and are never happier than when standing "on a jagged peak above the clouds, with the humdrum world forgotten and the stars for aristocratic company."[7] They are either like Plato (as Francis Bacon described him), men of sublime imagination "who take a view of everything as from a lofty rock"; or else they are men of incredible concentration, inclined to take refuge in their cloister, garret, laboratory or observatory, where they can shut out everything irrelevant and intensify their efforts on their peculiar studies, without dilution or distraction.

Montaigne once affirmed, with much truth: "We must reserve a little backshop, all our own, entirely free wherein to establish our true liberty and principal retreat and solitude."[8]

As a child, I used to hide behind and partly under a couch where I could become absorbed in drawing, writing, or reading – and lost in a world of my own. That was the first "little backshop" of my childhood. In recent years – once I could finally

---

[6] I invoke the hunter advisedly. It has been shown that moderately high levels enhance concentration and persistence. This is not only seen among hunters stalking game but also in males seeking desirable mates. See R. J. Andrew, Increased persistence of attention produced by testosterone, and it implications for the implications of sexual behavior. In: *Biological Determinants of Sexual Behavior*, ed. J. B. Hutchinson. Chichester, England: Wiley, 1978, pp. 255–275.

[7] As Christopher Benfey describes Percival Lowell in *The Great Wave: Gilded Age Misfits, Japanese Eccentrics, and the Opening of Old Japan* (New York: Random House, 2003), p. 177.

[8] Michel de Montaigne; *Essays*, Book I, ch. 39. It is more elegant in the French: "Il se faut réserver une arrière boutique toute notre."

afford it – I have found a better refuge in my observatory. Through the peephole of the eyepiece of my telescope I have projected myself millions of miles across space, in voyages as fantastic as any made by Aladdin. Within my hermit-like cloister, I could open the shutter and ascend, in mind, into the heavens. I was the high-priest in the Temple to the gods of my religion – the planets.[9] My observatory being located in a small field bordering a wood, I have felt like Thoreau there, and have often said with him:

> A slight sound at evening lifts me up by the ears, and makes life seem inexpressibly serene and grand. It may be in Uranus, or it may be in the shutter.[10]

The sounds are usually of crickets, which comfort the lonely observer through the long hours of the night; sometimes the mournful hoot of a solitary owl.

As a pre-teen, I participated in *epiphanies*, "showings" of the gods I worshipped; I enjoyed nightly what Percival Lowell called "revelation-peeps" of other worlds. By glimpses I came to understand that the ultimate reality is too deep ever to be known completely. There was always something of mystery left at the end, something unintelligible. *All the better!* Here was a very different perspective from that of the smug enforcers of religious dogma – the crushing and suffocating lead-blanket of certitudes – under which I had grown up.

A planet as seen in my small telescope was inconveniently minute; but what infinite riches in a little room! Venus showed a phase, Jupiter bands; that was all. Even so they suggested infinities. The first afforded a revelation of the Copernican system – of the Earth's place as a journeyman-world traveling around the Sun; the other granted a glimpse of cloudtops, the actual meteorology, of a globe much vaster than the Earth. Such things kept me wide awake for hours.

Above all, I waited for my first good look at Mars, the planet that was already my *chef d'oeuvre*. Its turn finally came in the order of things as it reached an unfavorable (aphelic) opposition in March 1965. I remember how "little it was, so silvery warm – a pin's-head of light!"[11] Odd, but I was not disappointed. I peered insatiably at that brilliant pinpoint, small in itself, vast in its implications. Ruskin says:

> The greatest thing a human soul ever does in this world is to *see* something, and tell what it *saw* in a plain way. Hundreds of people can talk for one who can think, but thousands think for one who can see. To see clearly is poetry, prophecy and religion, all in one.[12]

Mars did not then reveal to me the detail it would later when I had the chance to view it with ampler appliances. Yet even as little as it was, it had nevertheless vouchsafed something momentous and important and full of consequence to me.

---

[9]Mircea Eliade writes: "The most ancient sanctuaries were hypaethral or built with an aperture on the roof. – The "eye of the dome," symbolizing break-through from plane to plane, communication with the transcendent." In: *The Sacred and the Profane: the nature of religion*. Translated by Willard R. Trask. New York: Harcourt, Brace & World, 1959, p. 49.

[10]Henry David Thoreau, *Journal*, July 10–12, 1841.

[11]H. G. Wells, *The War of the Worlds*. In: *Seven Science Fiction Novels of H. G. Wells*. New York: Dover, no date, p. 312.

[12]John Ruskin, *Modern Painters*, Vol. III, Part IV, Chapter 16, Section 28.

It was a planet, another orb in space; it had a disk, and I was seeing it exactly as if from the porthole of a spacecraft only a million miles away. That was something. Mars I had *seen*, and *seen* in a plain way. There was poetry, prophecy, and religion for me in that.

Long before I had my own observatory, I set up my light and portable telescope wherever the light paths above rooftops or between the trees allowed a clear view of the object I was chasing, and observed *en plein air*. I did so in all seasons – even in winter, when Orion bestrode the next-door neighbor's porch (and winters in the 1960s were still cold in Minnesota; this was before Greenhouse Warming had registered an effect). But I found my greatest pleasure observing in the spring, when the air was warm and fragrant with the scents of flowers and crickets served as my musical accompaniment, and when to be young was very heaven.[13]

I was full of wonder then. "Wonder," said Francis Bacon, is "the seed of knowledge."[14] Spinoza, in his *Ethics* (Definitions of the Emotions, IV) calls it *admiratio*, "the conception (*imaginatio*) of anything, wherein the mind comes to a stand." The seed took root and grew, and my mind often came to a stand as I drank in beauty through the pipette of my telescope.

There was poetry in this – and, because my parents were indulgent, content to "leave me alone," my childhood, except for school, was on balance more poetry than prose. Perhaps there was something of the poet in me even then; something of what Wordsworth calls "Intimations of immortality" (but in *whom* are there not such intimations?).

When I left school for the day – tired, discouraged, or depressed, after being drummed at over tables of multiplication, diagramming sentences, and parroting religious dogma – I dreamed of other worlds and felt refreshed. My teachers were oblivious of my nocturnal trances; they never guessed by what means I was "lifted up when fallen."

I am willing to grant that there may have been something escapist in my drive to study the planets and stars. As a child I was naturally shy. As I said before, I was a loner, only too keenly aware that I was not like others in the things that were most meaningful to me. My most precious "spots of time" – for some reason Wordsworth keeps running through my head[15] – were the hours I spent with the planets and stars

---

[13]It is in the spring, I might add, Testosterone levels are highest. Needless to say there was (and is) something almost erotic about planets for me. I was especially drawn to the enchantress Venus, which in particular makes its most impressive showings (for Northern Hemisphere observers) in the spring, so that it could be said of me, as for most: "in the *spring a young man's fancy turns to love*."

[14]*Francis Bacon*, The Advancement of Learning, Book I.

[15]From *The Prelude* (1850), XII, 207. The whole passage is as follows:

> There are in our existence spots of time,
> That with distinct pre-eminence retain
> A renovating virtue, whence depressed
> By false opinion and contentious thought,
> Or aught of heavier or more deadly weight,
> In trivial occupations, and the round
> Of ordinary intercourse, our minds
> Are nourished and invisibly repaired.

or curled up with a favorite astronomy book; they gave shape and meaning and dignity to my life, they were a refuge as well as a passion.

I want to emphasize that astronomy, when my interest in it began, was a playful activity. My interest in it was self-driven, spontaneous, and occurred with rather dramatic abruptness – not unlike a volcano erupting. No adult – parent or teacher – put that puddle in my way as I returned home from school that day; none asked me to ponder the image of the Sun reflected in it. Had they done so, I doubt whether the result would have been the same or that I would have started playing around open-endedly with ideas about the Larger World that my imagination had suddenly annexed for itself. None of them gave me instructions to go out on spring evenings and follow Venus up and down the twilight sky. None of them put a telescope in my way. None attempted to structure my activity, as adults do through Church groups, Scouts, and sports teams. Had they done so, they would have spoiled it.

None of them sent me off to look for books on astronomy at the local branch library, or charged me with writing a book report or preparing for an exam. I discovered the resources I needed on my own, and put them to my own uses. Above all, once I got that telescope, however modest, I had my Aladdin's lamp, my magic carpet, my time-machine. Its tripod was like the stirrups of Pegasus; once mounted, I was borne aloft on feathered wings, I peered with eagle sight. Wielding that telescope was exciting, it was great fun. It opened up limitless possibilities to the imagination at the very moment when those were the things I wanted most. I was about to enter adolescence. Often seen as a problem-time, a period of tumult and alienation, Paul Shepard has written rather more sympathetically of the adolescent stage of life:

> A different view of the adolescent, less perplexing to adults and more nearly true, is to see him moving toward a peak of discovery and sensibility that he may never again experience. The extremes are his measuring of the range of life, the enormity of what it is possible to be like a landscape before his eyes, inseparable from what it is possible to feel and believe.[16]

In my case, the enormity of what it was possible to see, the landscape before my eyes, was cosmic in scope. And what I felt and believed was largely defined in terms of astronomy.

To the extent I was a loner, it was mainly by temperament and to some extent by choice; it was not because of exclusion or ostracism. I had my school friends – or at least friendly acquaintances; chums, allies, rivals. Many were ruffians. Some were shallow, some cruel; a few were deeper and more thoughtful. At the tasks that were set before me in school – the routine sessions where conformity always trumped creativity – I was never the best. I was studious nevertheless in my own time; but I was not a bookworm, and I was never the most laggard in running to the pond nearby where I would sail homemade boats guided by strings, or to the park to fly kites or model airplanes.

[16] Quoted in ibid., p. 151.

I differed from my friends only in that nightfall was a cue to me for something far more nourishing. It marked me again as a lonely one who would silently get my telescope out to begin what I considered my real work – adding not to baseball card or stamp or coin collections (though such pursuits absorbed me once) but to my growing stock of astronomical observations. I had always been an avid collector, and realize that collecting must be an instinct like the hording of nuts by squirrels in the fall or birds the bric-a-brac and yarn used to build their nests. Simon Garfield has said: "Collecting fills a hole in life, gives it a semblance of meaning; it is a diversion from reality – and others engaged in it form, together, a club of extreme enthusiasm. Collections help organize a world of chaos, and bring a dependable meaning to a life."[17] Whereas my peers went on collecting baseball cards or stamps or coins, and later trophies and cars and girls, I collected observations of planets, double stars, nebulae.

Parenthetically, I should mention that I was, as I had been earlier, mad about drawing, about recording everything I saw in minute detail, with notes about the times and circumstances of each observation. It was, insofar as I could manage it, representational, not abstract art; I tried to depict what I saw to the best of my ability. If I had lived in the Pleistocene, I undoubtedly would have covered the limestone walls of deep caves with the things that were most important in that world of the Ice-Age Mammoth Steppe – the Great Beasts. Living in the early Space Age, I drew the Moon and planets.

My collections – be they only scraps of paper with sketches and data about the times and circumstances of observations scrawled upon them – were extensions of myself. I made those records as if under compulsion; they were not done according to principle or plan, they were never organized or methodized. Ultimately, they expressed who I was. And is it not true, as someone – I think it was Jean Baudrillard – put it, "what you really collect is always yourself"? The Moon, the planets and stars were the things I loved and wanted to surround myself with. They were more than things; they were beings, personalities, alive. I could not say – nor did I ask – why I was drawn to them. It was a simple fact.

I hasten to add that my drawings were not mere spontaneous effusions of the moment; they were meant as records and, as poor as they may have been, they strove as art has always done to wage battle with time. I still have some of these drawings from 45 years ago. When I shuffle through them, they bring back my childhood, and they do for me what the 20,000 year old Beasts painted on the walls of Altamira did for Teilhard de Chardin. Teilhard found in them that "what we really discover is our own childhood, we discover ourselves, because we observe the same essential aspirations in the depths of our souls."[18] When I look at them, they don't seem like solemn or venerable documents; instead they seem like mere caprices. I still remember just how much fun I had doing them.

In my solitary hours with the telescope, I was master of all I surveyed: the entire Solar System – the whole universe of stars – might have been my own private mansion

---

[17] Simon Garfield, "The Passion that led me astray," *The Guardian*, March 30, 2008.

[18] Quoted in Guthrie, *The Nature of Paleolithic Art*, p. 115.

and estate. From the confining and paltry backyard of my parents' humble abode, I enjoyed – in the vertical direction – a view over territory so vast that it beggared the imagination. It was, to all practical intents and purposes, mine and mine alone. I tended to regard it so. The Moon and the planets and the stars never seemed to have other company; I was the only one present in the throne-room of these divinities, bearing the names of the classical gods and goddesses, when they granted their audiences with me – sometimes shrouding their majesty in incense and enhancing their mystery by playing hide and seek with me behind the clouds; now and then – sometimes best when the sky was hazy rather than brazenly clear, more often at twilight than in the deep night – the air grew steady and the markings sharpened like a steel engraving and confided to me and me alone a new revelation, an irregular twist to the Jovian bands, a dusky marking or polar cap on Venus, that made me wish for all the world for a larger telescope.

As I realize now, it was the intermittent reinforcement schedule that helped make planetary observing so addictive; it was impossible to know when to expect those supremely still and satisfying glimpses of fine detail. So I waited – and watched.

I was like a pagan priest born 2,000 years too late, worshipping gods that had not been worshipped, maybe, since Hadrian dedicated the Pantheon in Rome during the time of Ptolemy. I had my own rites, my own worship. I might have said to them as Keats did to the too-late born Psyche,

> So let me be thy choir, and make a moan
> Upon the midnight hours;
> Thy voice, thy lute, thy pipe, thy incense sweet
> From swingèd censer teeming.[19]

I have always admired eccentrics. They are the utterly committed ones – moving on a rim of their own, deviating from the rule, non-conformist and unconventional. I need hardly remind the reader that the term is ultimately astronomical; it refers to a circle that does not share the same center with other circles or, as in the ellipse, that deviates from the round, the circular, path. I was eccentric if not particularly deviant. Also, I believe, a disproportionate number of eccentrics are male; it may be a result of Testosterone, which causes a more asymmetric development of the two hemispheres of the human brain, so that men compared to women have a tendency toward obsession rather than balance.

Many, if not most, of the great astronomers were eccentrics, and while before February 17, 1964 I had had the usual heroes of one acculturated to masculine interests, such as Superman and the Lone Ranger, military or political figures, baseball and football players – thereafter my heroes were astronomers. They seemed to exist on an infinitely higher plane of being than those others. I recognized in them the same hearts touched by fire during the Dream Time of youth; they too had been as I felt now, "undefeated in spirit, invincible, pure." Their example taught me I was not alone. Their glory was the greatest glory – their names were inscribed forever among the eternal stars. They had aspired to be like gods, and at some level they were. Reading about them – thinking about them – they spurred me on.

[19] Keats, "Ode to Psyche," lines 44–47.

I admit that – dazzled as I was by what they had accomplished – I secretly believed, or hoped, that I might one day be counted among them, if I tried hard enough – if I did as they did. I too craved membership among the gods.

I read about Copernicus – Tycho – Kepler – Galileo. There was even a real astronomer in the neighborhood – Strathmore R. B. Cooke, a native New Zealander who taught geology at the University of Minnesota and was an accomplished observer of the Moon with a 12½-in. reflector which he housed in a shed in his backyard. He lived about 2 miles from me, and I often rode my bike to his house and regarded his observatory, from the alley, with envious eyes. I always hoped he might venture out, perhaps to empty the trash, when I was there, so I could introduce myself to him; but he never made an appearance, and I was too shy to knock at the door. We never met. It may have been just as well. I was decidedly green behind the ears at the time, and Cooke, like most serious amateurs, had a reputation for not suffering fools gladly.[20]

In high school, my ultimate hero was Isaac Newton, of whom it was said by one of his acquaintances at Cambridge:

Sir Isaac Newton. Profile of the ivory bust by David Marchand, now in the British Museum. It was described by Newton's half-nephew Benjamin Smith as being "from its elegance, similitude and placid expression truly valuable." It reputedly cost Newton a hundred guineas to have it made. Photograph by William Sheehan in 2009

[20] W. F. Denning, in his Notes on Telescopic Work, speaks discouragingly of "Friendly Indulgences." "Every man whose astronomical predilections are known, and who has a telescope of any size," he says, "is pestered with applications from friends and others who wish to view

"I never knew him to take any recreation or pastime either in riding out to take the air, walking, bowling, or any other exercise whatever, thinking all hours lost that were not spent in his studies."[21]

This was a man who burned the fire in his laboratory at all hours during the spring and fall, and kept watch in the quad at Trinity College observing comets over the roof of King's College chapel until he became "disordered in his senses." Newton was a man who experienced ecstasy – *total surrender* to a commanding interest, so that the rest of the world, even his own body and bodily needs, dropped away. He was the "Last of the Babylonians," according to the economist and Newtonophile John Maynard Keynes; a Shaman in a wig and tight-fitting hose. So, to greater or lesser degree, were they all.

I admired the Japanese amateurs Kaoru Ikeya and Tsutomo Seki, whose diligent hours sweeping the sky for comets led to the discovery of one that proved to be the most spectacular Sun-grazing comet of the twentieth century. (I remember reading a *Reader's Digest* article about Ikeya's superhuman dedication to comet-seeking, and it inspired me – in a spirit of emulation – to have my father wake me up in the dark before he went to work, so I could point my telescope to the planets in the morning sky.) Comet Ikeya-Seki blazed out in the predawn sky in October 1965; it was the first comet I ever saw.[22] While I was out looking at it, I also turned my 2¼-in. refractor at Jupiter, and had the first view of the planet in the brightening sky of the planet's belts – they appeared lovely reddish or coppery-hued against the yellow zones of that planet's gaseous globe.

I learned about the Great Comet from Robert Burnham, Jr., the first real astronomer I had contact with. I shall always be much in his debt. I was then captivated with the idea of searching for other worlds – to search is to hunt; there's the familiar

---

some of the wonders of the heavens. Of course it is the duty of all of us to encourage a laudable interest in the science.... but the utility of an observer constituting himself a showman, and sacrificing many valuable hours which might be spent in useful observations, may be seriously questioned. ... The time of our observers is altogether too valuable to be employed in this fashion... My own impression is that, except in special cases, the observer will best consult the interests of astronomy, as well as his own convenience and pleasure, by declining the character of showman; for depend upon it a person who appreciates the science in the right fashion will find ways and means to procure a telescope and gratify his tastes to the fullest capacity. Some years ago I took considerable trouble on several evenings in showing a variety of objects to a clerical friend, who expressed an intention to buy a telescope and devote his leisure to the science. I spent many hours in explanations &c.; but some weeks later my pupil informed me that he really could not afford to purchase instruments. Yet I found soon after that he afforded [a considerable sum] in a useless embellishment of the front of his residence, and it so disgusted me that I resolved to waste no more precious time in a similar way."

[21]Quoted in Frank E. Manuel, *A Portrait of Isaac Newton*. Cambridge, MA: Belknap Press at Harvard University Press, 1968, p. 104.

[22]This was not my first case of great good luck, astronomically speaking. I was also fortunate in having been present in the track of totality at the solar eclipse of June 30, 1954, which passed through Minneapolis. However, being only 12 days old, I am afraid I have no recollection of it!

Robert Burnham, Jr. Credit: William Sheehan collection

metaphor again. I had read a book describing Percival Lowell's mathematical quest for "Planet X" beyond Neptune, which after his death led to the discovery of Pluto by Clyde Tombaugh who had undertaken an exhaustive search with the 13-in. Abbott Lawrence Lowell wide-field refractor at Lowell Observatory.[23] (Pluto was, of course, then classified as a planet. It has since been demoted, but I will always think of it as a planet; perhaps eventually the International Astronomical Union will change its collective mind and agree.)

[23] The book was fine juvenile offering: Tony Simon, *The Search for Planet X* (New York: Basic Books, 1962). I returned to it recently, and found I could still read it with pleasure.

The 13-in. astrograph used by Clyde Tombaugh to search for Percival Lowell's "Planet X," which led to the discovery of Pluto in 1930. Photography by William Sheehan in 2004

I dreamed of conducting my own search for planets in trans-Neptunian space. One day I was "blinking" two images of Pluto reproduced in James Jeans's book *The Universe Around Us*. (I did not have a blink comparator, of course, but did the job economy-style; I alternated closing and opening my eyes in quick succession. But the principle – the parallax effect – was the same.) The images had been taken by C.O. Lampland at Lowell Observatory in the aftermath of Tombaugh's discovery, and showed the planet passing close to the star Delta Geminorum (which appeared vastly overexposed with rays from the diagonal vanes in the left lower corner of the images). Suddenly, I found something: a tiny speck seemed to be jumping near that star, and I convinced myself from its rate of motion that it might lie at perhaps twice the distance from the Sun as Pluto itself. At once I wrote to "Lowell Observatory" about it. With a heavy dose of make-believe and suggestibility characteristic of a child of that age, I fantasized having discovered a planet – one all the great astronomers before me had missed! – and grandiosely imagined being invited to become a staff astronomer at Lowell at the tender age of 11!

I received a kindly answer from Robert Burnham, Jr., "Bob" as he signed it. He had begun as an amateur astronomer himself. After discovering six comets with a homemade telescope and attracting the attention even of the prominent Arizona Senator, Barry Goldwater, he had been invited to join the staff at Lowell, where he was assigned to the "proper motion" survey – his job was to take a fresh set of plates with the "Pluto" telescope and compare them with those of the same star fields by Tombaugh 30 years before, in order to identify stars with large proper motions. That was his official role. His unofficial role was as an encourager of children with astronomical interests and – rather than being put off by my obvious grandiosity and ignoring such a letter as almost anyone else in his position would have done – he took time to type out a very kind, one-page, single-space letter on vellum paper in which he mentioned that – since the Lowell library did not have a copy of Jeans's book – he had gone to the trouble of going to the plate vault and searching out the original plates exposed by Tombaugh and reblinking them. Since he did not find my planet, he suggested that dust or other flaws introduced in the process of preparing the images for publication had probably been responsible – my "planet" was nothing more than a flaw in the plate or a speck of dust. But instead of ending on that distinctly discouraging note, he gave me pointers about observing the "Great Sungrazing Comet" of 1965. Thanks to his letter – and realizing I had registered on the consciousness of a real astronomer – I was over the Moon for days with excitement. Moreover, Bob Burnham's instructions helped me spot the Comet in the pre-dawn sky as it made a hairpin turn around the Sun, streaming its trail of glory across a span of some 120°. My memory of Bob Burnham will be forever entwined with my memory of that Comet.

From what I learned later, the fact of his writing to me seems even more providential and generous than it seemed at the time. Bob Burnham was – as later emerged – a very shy person, to put it mildly.[24] Even at Lowell he spoke rarely to anyone. His niece who sometimes came from Prescott to visit him would often find him sitting in a rocking chair on pine needles outside his cabin, enjoying the silence of that magic place I have since come to know so well: inside the cabin was a fascinating clutter of rocks glowing under ultraviolet light, ancient coins, other artifacts of long-dead cultures, fossils of trilobites and sharks' teeth, and on every wall, from floor to ceiling, *books*. His mind was similarly stuffed with information, and at the time he wrote to me, he was busily mimeographing and assembling the pages of the voluminous masterpiece which became the *Celestial Handbook* (originally self-published, it was later issued by Dover Publications, to which he sold the rights for a flat fee of $2,500). The observatory was his refuge; later – when the funding dried up for his research project and he was let go – he sank quickly out of sight, rather like a comet receding from the Sun. In the end, he became a street-person who tried to live off the money he received for paintings of psychedelic cats in a park in San Diego and died in 1993, his health broken, intermittently psychotic, depressed and neglected, in a flop house.

Burnham's fate underscores another thing about our ilk. Many of us aren't well suited to anything practical or useful. We're harmless creatures; but we're not really good for anything – except to sit in rocking chairs on pine needles and hoard curiosities (artifactual or informational) that to most people will seem mere bric-a-brac – cast-offs – debris. We are beachcombers along unfamiliar seas who love to peer, whenever we can, at shining planets and spangled multitudes rolling through the vast universe of which we, the meek, are the true inheritors. Those planets,

[24] For information about Burnham, I am following Tony Ortega, "Sky Writer," *Phoenix New Times*, September 25, 1997, which is likely to remain the definitive account of Burnham's sad life.

which are a mere backdrop – distant, unimportant – for most of the human race, constitute for us – as George Steiner said of the addict to chess – "a reality, a focus for the emotions, as substantial as, often more substantial than, reality itself."[25] We pursue our devotion despite the fact that the financial element, where it exists at all, has always been small or accidental.

I would, in my current capacity as a professional psychiatrist, estimate conservatively that at least a quarter and possibly half of all the really single-minded amateur astronomers I have known have had some version of Asperger's Syndrome and suffer, in high degree, from what might be popularly described as "nerdiness." I don't say this in some superior manner; I include myself among them. Though I've rounded out a bit socially since my early adolescent days, I can still remember when – though I lacked the coke-bottle eyeglasses – I would have stood out in most crowds as a "little professor" (a nickname to which Burnham was also subjected). I am someone with unusual interests – a social misfit and a loner.

I do not doubt that I was regarded as an eccentric by most of my more conformist and ordinary companions in those days – and if so, what they supposed was no more than the truth. Though the whole experience was not without its painful aspects, I tried to regard the difference as a badge of honor and, where once I had hoped to be a figure of action like the baseball players and football players I admired in earlier years, by the age of 10 or 11 I had found my true self, and preferred men of thought over men of action. My role models were brainy isolates like Newton, Henry Cavendish, Burnham, or H.G. Wells's Cavor (who built the Cavorite sphere in the *First Men in the Moon*; my clinical training allows me to easily recognize that this fictional character suffered from Tourette's syndrome). Wells deliciously describes Bedford's first glimpse of Cavor out walking:

> The sun had set, the sky was a vivid tranquility of green and yellow, and against that he came out black, the oddest little figure.
>
> He was a short, round-bodied, thin-legged little man, with a jerky quality in his motions; he had seen fit to clothe his extraordinary mind in a cricket cap, an overcoat, and cycling knickerbockers and stockings.... He gesticulated with his hands and arms and jerked his head about and buzzed. He buzzed like something electric. You never heard such buzzing. And ever and again he cleared his throat with a most extraordinary noise.[26]

England – because of its class system – has been famous for its eccentrics, cranks, and borderline nutters; the idea being, presumably, that it's acceptable for aristocrats to be eccentric, and even a badge of honor. People who are well-to-do and privileged can do just as they like. When I started out, amateur studies of the Moon and planets had for decades been dominated by a long succession of "mad Englishmen" dating back to the early days of the British Astronomical Association (founded in 1890). Their ranks included Arthur Stanley Williams, who lived in a boat and had his telescope on shore; the stage and screen comedian W. T. Hay, whose most famous role was as an eccentric schoolmaster and who discovered the

---

[25]George Steiner, "A Death of Kings." In: *George Steiner: a Reader* (New York: Oxford University Press, 1984), p. 173.

[26]H. G. Wells, *The First Men in the Moon*, op. cit., p. 8.

Great White Spot of Saturn of 1933; the famously eccentric Patrick Moore, who with his monocle and xylophone and dogmatic opinions could not fail to make a very strong impression.

Many an amateur I have known has resembled William Frederick Denning, a famous late-nineteenth century member of the B.A.A. of whom Richard Baum has written:

> Several days ago J. Hedley Robinson phoned me… the last director of the B.A.A. Mercury and Venus Section … and spoke about that great meteor observer W. F. Denning. Since Hedley … lived in the vicinity (Bristol), I wondered if he had ever met the man. "No," he replied, "he seemed a bit aloof." Apparently Denning lived alone with his man servant – he was an accountant by profession. I was also told, some years ago, and the letter is still in my files, that the late Dr. W. H. Steavenson, a very well known English amateur, once visited Denning, to discover a lonely old man sitting by his fireside. James Muirden, author of *The Amateur Astronomer's Handbook*, told me that tale; he also added that his (James's) father had lived a few streets away from Denning and could well imagine how the street urchins used to catcall and abuse him as he made his way home.[27]

There he was, still alone; now by his fire as once by his telescope.

As eccentric as these individuals were, they had nothing on Giovanni Schiaparelli, the great nineteenth century Italian observer of the planets, renowned especially for his studies of Mercury and Mars. He was one of a small pantheon of

Giovanni Virginio Schiaparelli. William Sheehan collection

[27]Richard Baum to William Sheehan; personal correspondence, November 24, 1989.

astronomical deities I worshipped as a youngster (which is to say I took them as role models, aspired to be like them). He was, apparently, absent-minded in a degree that threw even Isaac Newton into the shade. As his niece Else Schiaparelli, the famous fashion designer, recalled in her autobiography:

> He was appallingly absent-minded. After his marriage he took his young bride to Vienna. On the evening of their arrival he exclaimed:
>
> "Excuse me, I must go and see an astronomer who lives in this town. I won't be long."
>
> He rushed off with his mind already full of telescopes and stars while his poor little wife stayed sobbing in the hotel bedroom. She waited for dinner. He did not come back. Midnight struck. And then the whole night went by without any sign of Giovanni. The next morning my uncle, quite unconcerned, returned to the hotel, asked for the key, and went up. There on the floor, a pathetic bundle, was his sweet wife still sobbing.
>
> The astronomer let out a cry of surprise.
>
> "Oh!" he exclaimed, running to her. "I completely forgot I was married."[28]

A few astronomers are social (though even then they like to associate with others of their kind – thus the interesting human fare one encounters at star parties and astronomical meetings, which contain a high preponderance of "nerds." They enjoy talking about their passion – enthusiastic about that which most of the rest of the world holds in indifference – and shut out that world with a kind of code. To outsiders, such gatherings may well seem reminiscent of the Bar Scene in *Star Wars*). Most of us admittedly are loners to a degree. Denning and Burnham obviously are good examples. Burnham, for instance, interviewed himself – a very autistic thing to do – for *Astronomy* magazine in the early 1980s; it was a very strange interview. Part of what he said was: "We're all a lot crazier than we think. Though we show it in different ways. I'm a virtual hermit, for example, and never attend astronomical meetings."

I suspect most of us are set apart even from childhood. I resonate – and I suspect most dyed-in-the-wool amateur astronomers would – with the words of Daniel Tammet, synaesthete-extraordinare, who is noted for setting a record in memorizing pi to the most decimal places (22,514). Tammet sums up his experiences in the nursery:

> When the time sometimes came to play social games, such as musical chairs, I refused to join in. I was frightened by the thought of the other children touching me as they shoved one another for one of the remaining seats. No amount of gentle persuasion by the supervisors would work. Instead I was allowed to stand by one of the walls and watch the other children play. So long as I was left to myself I was happy.
>
> The moment I came home from the nursery I would always go upstairs to my room...
>
> My room was my sanctuary, my personal space where I felt most comfortable and happy. I spent so much of my day there that my parents took to coming up and sitting with me in order to spend time with me....[29]

---

[28]Elsa Schiaparelli, *Shocking Life*. New York: E.P. Dutton, 1954, p. 25.

[29]Daniel Tammet, who in *Born on a Blue Day: a memoir*. New York: Free Press, 2006, p. 27.

Tammet's room, Burnham's cabin with its antiquities, the Cavorite sphere, an astronomical observatory – all of these are variations on a common theme. From an early age I wanted to have my own private place where I could retreat – safely – from human contact and enjoy something more objective and real than the world of social relations or merely human things. I liked the cozy womblike confines of a close and sheltered little space. From the security within – and without fear or anxiety – I could open the shutters and look into the universe beyond and so enjoy the best of both worlds. That basic formula – a close secure (and usually crowded and disorderly) space all my own from which to launch far-flung adventures has remained with me throughout my life. I prefer the cottage to the palace; care to have no mansion other than that of my own restless and vagabondish mind. As for a place to house the body, a Thoreau-like cabin on some Walden Pond will do. In my case – as for Thoreau and for many an amateur astronomer else – Walden Pond is the universe, no less.

We need nothing more than a spot from which to look out with wonder, some place from which to throw our gaze at a picture-window-view far grander than that from the most extravagant Great House or Palace, with lawns and woods and territory so vast that imagination cannot compass it. We need some place like Ogilvy's observatory in the opening chapter of *The War of the Worlds*:

> the black and silent observatory, the shadowed lantern throwing a feeble glow upon the floor in the corner, the steady ticking of the clockwork of the telescope, the little slit in the roof – an oblong profundity with the star-dust streaked across it. Ogilvy moved about, invisible but audible. Looking through the telescope, one saw a circle of deep blue and the little round planet swimming in the field.[30]

I have said that astronomers, as a whole, are awkward for general purposes, aloof, shy, unsociable, or simply engrossed. But I would argue that they have all, at one time or other, experienced something kindred to what the painter John Mallard William Turner experienced:

> At last fortune wills that the lad's true life shall begin: and one summer's evening, after various wonderful stage-coach experiences on the North Road... he finds himself sitting alone among the Yorkshire hills. For the first time, the silence of nature around him, her freedom sealed to him, her glory opened to him. Peace at last; no roll of cart-wheel, nor matter of sullen voices in the back shop; but curlew-cry in space of heaven, and welling of bell-toned streamlet by its shadowy rock. Freedom at last. Dead-wall, dank railway, fenced field, all passed away like the dream of a prisoner; and behold, far as foot or eye can race or range, the moor, and cloud. Loveliness at last. It is here, then, among these deserted vales! Not among men. Those pale, poverty-struck, or cruel faces; – that multitudinous, maimed humanity – are not the only things that God has made. Here is something that He has made which no one has marred. Pride of purple rocks, and river pools of blue, and tender wilderness of glittering vales, and misty lights of evening on immeasurable hills.[31]

[30] Wells, *The War of the Worlds*, p. 312. I have always admired poor Ogilvy, though he is a character in fiction; he perished in the pit in which the Martians landed on Horsell Common. I almost named my observatory after him.

[31] John Ruskin, "The Two Boyhoods," in *Modern Painters*, volume V, part 9, chapter 9.

He might have added for good measure a brilliant planet or two – and rafts and rafts of stars.

What Turner realized Percival Lowell came across as a 3-year old boy climbing the winding staircase of the family mansion located on Boston Common and being presented, through the window, with the graceful scimitar of Donati's Comet arching across the sky. E. E. Barnard – born into less magnificent circumstances – realized it as a small boy from a wagon-bed in the slums of Nashville during the sad days of the Civil War. Similar moments of revelation are found in the biographies of most men of genius. Most of them also come face to face with a terrible loneliness at one time or other. Eventually they make of their limitations a virtue. They make their home in the waste places – the solitudes, wildernesses, and deserts. They all come, sooner or later – usually sooner – to realize, as Turner did, that "humans are not the only things that God has made." Nor perhaps even the best.

Comet Donati, as it appeared on October 5, 1858. The head is situated near the bright star Arcturus, in the constellation Bootes, and the scimitar-like tail, widely remarked, made this one of the most beautiful comets on record. From: E. Weiss, *Bideratlas der Sternenwelt*, 1888

When I was in high school, I remained an avid astronomer – and by then had stepped up to a 4¼-in. reflector, which first showed me the polar cap and dark markings of Mars. But I also became a devotee of distance running; something about which, for a few years, I was as passionate as I was about my astronomy. At first I shared

Young E.E. Barnard, aged about nine, posing as Oliver Twist, with John Van Stavoren his employer and the owner of the Photograph Gallery in Nashville, Tennessee. Credit: William Sheehan collection

the sentiment of poor Charles Hamilton Sorley (promising poet; his life ended, at 20, with a sniper's shot to the head at the Battle of Loos, France, 1915):

> We run because we like it
> Over the broad bright land.

Later I began to take myself more seriously, and eventually worked myself up to become a decent miler.

Running, I realize, satisfied me because it was something that appealed to my lonely, aloof and individualistic temperament. I was not inclined to be part of a pack, much less the conforming herd; "team sports," especially contact sports, with their quasi-military air – their love of uniforms and displays and parades and exhibitions of mimic fight – I had once loved but now found aversive. Being around others who thrilled to such only made me more keenly aware of my own singularity.

In running I could be myself. I ran for the same reasons that others have run – others who have done so for millions of years and who, perhaps figuratively rather than literally these days, still do run. (At night they chase the antelope – as I do the planets – in their dreams.) As runner and biologist Bernd Heinrich says:

> The human experience is populated with dreams and aspirations. For me, the animal totem for these dreams is the antelope, swift, strong, and elusive. Most of us chase after "antelopes," and sometimes we catch them. Often we don't. But why do we bother to try? I think it is because without dream-"antelopes" to chase we become what a lapdog is to a wolf.[32]

For me in my teen-aged years, running – like observing – was a quest for "game," and it involved some of the same rewards. (Food was, as Heinrich points out, always a distant, secondary objective; the hunt is prize enough in itself. As Sorley says: "We do not run for prize.") It also gave me the status of an athlete, which helped me to survive in a social system – typically American – where intellectuals are only too often harassed and persecuted.

My running, like my observing, allowed me a space, a void that served a necessary encounter with nothingness that balances the verbose, the prosaic side of life. Later I began to get into mountain climbing – or more accurately, hiking. I enjoyed the solitudes of open spaces – the magnificent – unfettered – ungirt views of the heights. Skies as far as eye can see. More voids, more balances to the verbose and prosaic.

Standing on a mountaintop, I could in a manner of speaking look "down" on the Earth. But the Earth itself – and any point therein from which one enjoyed an unobstructed view of the skies – was a mountaintop of sorts from which to look "down" on the entire universe. It was easy enough to learn the trick of reversing the perspective as I'd learned to do on that February day when I went splashing homeward from school and saw the Sun as a star and the stars as other suns. That *eureka* moment was in some ways the turning-point of my life: henceforth the Earth was a perch, a lofty vantage-point; not a place oppressive, subordinate, forever below. From here, I could look down on the whole rest of the universe and see shrink to mere points of light the other globes of space. And it followed that someone standing where *they* were – and looking this way – would see the Earth as a shining circle and finally as a mere point. All of this in which we are so wrapped up seemed, by that measure, dross.

Our world is a point; then what are we? I had a nun in second grade who used to say: "You're not such a much." The universe does not extol the grandness of man – but comments with eloquent silence upon his insignificance. At some level the burning away of our own ego is purifying. To contemplate the heavens is to participate in a kind of cosmic Zen. To participate in such vastness is – at least briefly – to annihilate the troublesome sense of self and its problems – to achieve Nirvana by becoming, by definition, part of something infinitely grander than one's self (even if only for a few moments).

As one who was "long in city pent," and who grew up in rather homely surroundings in a rather squalid blue-collar neighborhood that was fast becoming an inner city slum even during my childhood and adolescence, much of what there was of beauty in my life was for me unfolded in that theater of endless variety

[32] Bernd Heinrich, *Racing the Antelope: what animals can teach us about running and life*. New York: HarperCollins, 2001, p. x.

that played overhead in the sky (and one could still see the Milky Way from there; something that has been impossible, because of light pollution, for at least 40 years).

In a sense, our everyday lives – if only figuratively – too often resemble trench warfare. That being the case, the sight of the sky – almost alone – has the power to raise the mind from despair. As John Keats once said, "the world is full of troubles and I have not much reason to think myself pestered with many… I scarcely remember counting upon any Happiness – I look not for it if it be not in the present Moment… The setting sun will always set me to rights."[33] The setting Sun, the Evening Star.

Those in the trenches in World War I, Paul Fussell has pointed out,

> experienced an unreal, unforgettable enclosure and constraint, as well as a sense of being unoriented and lost. One saw two things only: the walls of an unlocalized, undifferentiated earth and sky above…. As the only visible theater of variety, the sky becomes all-important. It was the sight of the sky, almost alone, that had the power to persuade a man that he was not already lost in a common grave.[34]

Siegfried Sassoon recalls thinking once at evening stand-to that "the sky was one of the redeeming features of the war." For me, it was one of the redeeming features of childhood.

Though the sky can be discovered from the trenches or the inner city – and is perhaps "counted sweetest" by those who have so little else of natural beauty around them – the Open Sky is only found far from congested human habitations. Perhaps because until the Industrial Revolution, most of England was still rural – and the Open Sky was something too omnipresent to be noted or especially prized – it was not until John Ruskin published "Of the Open Sky" in the first volume of *Modern Painters* that sky-awareness became something associated with moral benefits – with the ability to speak to the human heart and to sooth it and purify it "from its dross and dust."

It is the grandeur of this sky, which the nineteenth century first discovered, that makes deserts and plains appealing, that draws some of us to the desolate grass-lands of the Dakotas, to the Big Sky Country of Montana, to the deserts of Arizona. Without the sky, they would be intolerable. The great writers about the desert are inevitably great writers about the sky; John Van Dyke, for instance.

I skip ahead to 1982, another pivotal year for me. After college, I drifted for a few years during which the one thing that sustained me was amateur astronomy. That summer I spent 2 weeks doing historical research on Mars at the Lowell Observatory – work that eventually led to the writing of a book, *Planets and Perception* – and then set out with a friend on a pilgrimage through British Columbia, Alberta, and Montana. Finally we encamped on the side of Chief Mountain – Mt. Ninaistakis, as the Blackfeet Indians (to whom it was sacred) called it – looking at the Grinnell Glacier to the west and across the plains and wheat fields

---

[33] John Keats to Benjamin Bailey, November 22, 1817.

[34] Paul Fussell, *The Great War and Modern Memory*. London: Oxford University Press, 1975, p. 51.

of Alberta to the north. A late afternoon thunderstorm had skirted around us as it followed the ridge to the west, but it didn't bother us; the sky filled with myriad cloud forms, and as the Sun set and threw its rosy splendor over the mountains, the lights of distant towns, some 50, others perhaps a hundred miles away, appeared on the plain below us as so many star clusters. That was but a prologue to the real drama of the stars appearing above, which shone with a brightness perhaps little undimmed from the days when before the White Men came. I do not believe I had ever been (except for the presence of my companion) so far from other human beings up to that time; when we at last turned in, and the wind rose in fury in the darkness, threatening to hurl our tent off the ridge, I was unable to sleep – not only because of the slapping and cracking of the canvas, but because I could not contain the excitement of what I had discovered about myself.

I had discovered that I have the heart of the hunter-nomad.[35] My habitat is the plains, the mountains, the wilderness, not the city or the farm or the club. As I am nowhere more at home than under John Ruskin's Open Sky, in my professional career I have always opted to work far from the Cities – out in the wide spaces – ruralized to obscure places in western Minnesota, North Dakota, New Zealand. I have worked in places where I enjoyed dramatic vistas of the skies, where I could explore new frontiers with my telescopes.

For the past more than 30 years I have continued to observe. There was a hiatus of a few years between high school and the end of college. My first telescopes had been stolen from my parents' garage shortly after the Great Dust Storm on Mars of 1971, when I was still in high school; being of limited means, it was several years before I could afford a replacement instrument. Instrument-making, alas, has never been my forte, and yet after 1982 – when I once more committed myself to serious observing – I had no choice but to concern myself with equipment and technique. Though I aspired to Maslovian "peak" experiences, sometimes literally, and have come to realize that I am a panoramic Big-Picture sort of person, I had to be satisfied that all my gear was in order. Observing is in that way like serious mountain-climbing – it is exhilarating, the rewards are great and, though not physically as hazardous, the pitfalls are many. One either needs to learn these things for oneself, or find others who are meticulously, obsessively attentive to detail, who are pedantic students of technique. Robert Graves, the poet and novelist, took up mountain climbing after the First World War, and found such a one in Geoffrey Young,

[35] I assume that means I have the seven-repeat (7R) allele of the human dopamine receptor D4 (DRD4) gene which has been associated with both attention-deficit/hyperactivity disorder and the personality trait of novelty-seeking which, though proximally I can trace it to inheritance from my alcoholic and vagabonding maternal grandfather, William Robinson, is a mutation that seems to have occurred quite recently during human evolution (during the late Pleistocene) and been strongly selected for as conferring some evolutionary advantage during those climatically unstable and wild times. See: Y.-C. Ding et al., Evidence of positive selection acting at the human dopamine receptor D4 gene locus, *Proceedings of the National Academy of Sciences* (January 8, 2002), vol. 99, no. 1, 309–314. For an interesting perspective on ADHD, see Thom Hoffman, *Attention Deficit Disorder: a different perception* (Grass Valley, CA: Underwood Books, 2nd ed., 1997).

an Eton Master and President of the Climbers' Club. Young had lost his four closest friends to climbing accidents. That Young survived was a testimony to his own cautious and meticulous nature. Graves described "the extraordinary care which he always took… It appeared not merely in his preparations for ascent – the careful examination, strand by strand, of the Alpine rope, the attention to boot-nails and the balanced loading of his rucksack – but also in his caution on the rock-face. Before making any move he thought it out foot by foot, as though it were a chess problem."[36]

Most mountain-climbers soon learn that – at least on the large peaks – a good deal of an expedition is involved with preparing one's equipment and carrying loads between the base camps. Climbing a mountain has often been compared to a military siege and, as Mick Conefrey has pointed out, "mountaineering is by its very nature a waiting game and most climbers accept that on a big mountain there is always a lot of 'pit-time' when bad weather keeps them trapped in their sleeping bags."[37]

Even with the telescope carefully collimated and in working order, there's still the climb. Before the spacecraft era, planetary astronomy was entirely ground-based. That meant trying to capture a column of light from a distant world by means of a lens or mirror then studying it with the eye, the photographic plate, or the spectroscope. Practically speaking, this meant a team collaboration among artisans – craftsmen in glass and mechanical engineers – who produced the instruments and the artists, trained in the methods of observation, who used them.

At one level, the classic pose was set in 1910 by Percival Lowell, hiding his bald pate under a golf cap turned with the visor backward peering intently at another world (in this case, Venus against the daylight sky). One of the challenges of mountain-climbing is that, though the mountain stands quite still, it is ever-changing, subject to the constant vicissitudes (and perils) of weather. The planetary observer's challenge is not with weather as such but with "seeing."

No telescope can ultimately perform better than the limit set by diffraction, which is a consequence of the wave nature of light. Owing to diffraction, a telescope can never form the image of a star as a perfect point; instead, the image consists of a small disk (called the Airy disk) surrounded by a series of bright and dark rings. The larger the telescope, the smaller the apparent disk and the more closely spaced the rings. Diffraction also determines what can be seen on a planetary surface; for instance, a thin line on a planet's surface is widened by diffraction into a band whose intensity shades off on either side. If a line is below a certain width, the contrast with the background is so much reduced that the eye is unable to grasp the weakened tones.

The diffraction limit to resolution is invincible. The only way around it is to increase the aperture of the telescope being used. However, as long as we are abyssal

---

[36]Robert Graves, *Good-bye to All That*. Garden City, NY: Doubleday, 1957, p. 63.

[37]Mick Conefrey, *A Teacup in a Storm: an explorer's guide to life*. (New York: HarperCollins, 2005), p. 90.

Percival Lowell observing Venus with the 24-in. Clark refractor at Lowell Observatory. Courtesy of Antoinette Beiser, Lowell Observatory

creatures living on the surface of the Earth, which is actually the deep sea floor beneath a massive ocean of air, this is always only a partial solution, for air waves interpose their own barrier to seeing. The column of light captured by the telescope (and ultimately directed to the eye) bears with it an adventurous history: after starting in the Sun, crossing interplanetary space between the Sun and the planet, reflecting off the planet's surface, then crossing interplanetary space again between the planet and the Earth, it must embark on the most hazardous part of its journey – the transversal of that roiling and tempestuous ocean of air. The light ray

> Into the wild expanse, and through the shock
> Of fighting Elements, on all sides round
> Environ'd wins his way; harder beset
> And more endanger'd, than when Argo pass'd
> Through Bosporus betwixt the jostling Rocks:
> Or when Ulysses on the Larboard shunn'd
> Charydis, and by th'other whirlpool steer'd.[38]

It arrives – if the reader will pardon another Miltonism – "like a weather-beaten Vessel … Shrouds and Tackle torn."[39]

---

[38] John Milton, *Paradise Lost*, II, 1013–1020.

[39] Ibid., II, 1043–1044.

The greater the telescope, the wider the captured column of light and the larger and stronger the areas of atmospheric turbulence through which it has had to force its way. With the small telescopes I used in my early years, I rarely had to worry about "seeing." I was limited by diffraction not by the roiling atmosphere (the downside was that I couldn't see very much). But as one increases telescopic aperture, one finds that the Earth's atmosphere rarely allows a telescope above a certain size – say 12 or 16 in. (30 or 40 cm) – to be used to advantage, at least by a visual observer.

The moving atmosphere causes the planet to appear to be in motion. It never – or hardly ever – stands completely still. There is an actual dancing around of the orb from its mean position in the field of view so that, as some amateur once quipped, in good seeing the images seems to be doing a waltz, in bad seeing a jitterbug. There is also a slower pulsating of the image that causes it to appear more or less blurred and "soft" until suddenly and without warning there occurs a flash of seeing in which the image turns sharp as a steel-engraving. These effects constitute "seeing" in the technical astronomical sense: the condition of the atmosphere that causes the planetary observer to be witness to something kindred to a dance of the seven veils. The observer is forced to be constantly be on the *qui vivre* and opportunistic: discarding the moments of indifferent seeing but ready to mentally seize what is vouchsafed during a sudden "revelation peep" – a moment of revelation in which the usually ruffled diaphanous veil of the Earth's atmosphere smoothes and straightens out.

It is no exaggeration to say that the observer resembles a hunter on the prowl who awaits his opportunity to strike. The game is afoot. In the case of the observer, the game is fine planetary detail. There is something exhilarating in the pursuit – an underlying current of excitement in the nervous system, something that tickles some gene defining a trait that must have appeared very early in human evolution and – I suspect – is related to what natural history writer R. Dale Guthrie calls the "appetite for hunting" which has characterized humans since the Pleistocene and beyond:

> In the evolution of hunting the need for nutrition is reinforced not only by an appetite for meat but by a thirst for the jubilation of the seek-and-kill experience. Though carnivores evolved a deep attraction to the taste of meat, the forces that drove hunting did not stop there. Their food does not hold still; it hides and fights back. Therefore, in addition to the delicious taste of red meat, natural selection added a predilection to hunt in the form of a special passion – which takes very little experience to activate. Among carnivores, we can see that individuals who have this passion perform better than their compatriots who hunt for hunger alone. And hence virtually no carnivores are without this ardor.[40]

In the case of the planetary observer, the pursuit of the game is just as intense and no less thrilling. The planet too has its bulwark of defenses, its forms of camouflage; it does not simply stand to be taken but "hides and fights back." When it is finally

[40] R. Dale Guthrie, *The Nature of Paleolithic Art*, pp. 246–247.

taken, the exhilaration is like that hunters feel in killing a Great Beast. The astronomer's quest is for intellectual nourishment and ends without bloodshed (unless, that is, the mosquitoes happen to be biting). On the whole, it seems likely that the passion for the planets among those who study them with the intensity of monomania is a much-sublimated form of the timeless passion to hunt.

When we look at a planet – when, for that matter, we look at *anything* – what we actually see is "a staccato of fixed images; each immediately erased by the subsequent one."[41] Somewhere, someone compared observing a planet to watching a movie with the projector out of focus except for brief, random intervals in which one or a few sharp frames occur.[42] A trick of the brain called flicker-fusion takes these static images and fuses them to make them appear to be a seamless continuum (which makes cinema possible). Thus, says Guthrie:

> We actually see static images, which our brain compiles into the process of movement, like the frames of a movie, only we are unable to "stop action." We are not able to isolate a momentary position within a moving series. It is much easier to recall or construct a mental image of someone standing still than it is to imagine that person in motion.[43]

In the jargon of CCD (charge coupled device) imaging, the eye resembles a digital device with a capture rate of 5–15 images per second. That means that all sorts of things in motion are impossible for us to see correctly. Though horses and other quadrupeds must have been objects of the most intense scrutiny by humans since time immemorial, the detailed action of a horse's trot was not worked out until the nineteenth century when Eadweard Muybridge used stop-action photography to settle a bet. A planet is no less an object in motion which the human eye-brain-hand system strains to freeze-frame. It is this fact that explains the difficulty of the pursuit, while the "fascination with what's difficult" accounts for why some few of us have found it so perennially absorbing.

I would add the obvious point that the observer is not subject to a regular reinforcement scheme of fixed rewards but to the intermittent one that is most apt to prove addictive. A first splendid look through the telescope can lead to higher and higher stakes viewing, and a continuous process of upping the ante through longer and longer vigils at the eyepiece. Like all addictions, this one can sometimes get out of hand, reaching the point where one's relationships and job begin to suffer. One seeks larger and larger instruments, lays out greater and greater sums to travel to remote places – usually to mountaintops – all in pursuit of bigger game and ever more gratifying rewards. Add to all that the chance of a *real*

---

[41] Ibid., p. 99.

[42] William Sheehan, *Planets and Perception: telescopic views and interpretations*. Tucson: University of Arizona Press, 1988, p. 99. In the same work, I also coined the term "tachistoscope effect," since the effect is similar to what a viewer would see through this device, used by perceptual psychologists at the turn of the twentieth century to analyze perceptions during brief exposures.

[43] Guthrie, *The Nature of Paleolithic Art*, pp. 99–100.

discovery – those who have tasted it, even if it is only something modest and of marginal scientific importance, such as the detection of some new disturbance on Jupiter or the visualization of a flare on Mars – will never forget it. The thrill then is one of the purest, most intense pleasures that humans are capable of, in which every moment – every chance association of scene and circumstance – are stamped irrevocably into the nervous system. No one has captured that feeling better than Keats in his famous passage from "On First Looking into Chapman's Homer":

> Then felt I like some watcher of the skies
> When a new planet swims into his ken,
> Or like stout Cortez, when with eagle eyes
> He stared at the Pacific; and all his men
> Looked with each other with a wild surmise,
> Silent, upon a peak in Darien.

Though I was long devoted to that pursuit, I must admit that even I have found the interest in visually observing the planets waning somewhat in recent years and giving way to the new pursuit of CCD imaging. Though the human eye remained the supreme arbiter of planetary detail for a full century after it had yielded pride of place to the photographic plate in the study of stars and nebulae, the eye has in the past 20 years or so been roundly surpassed by video cameras which capture frames at an even higher rate.

I confess that as someone who has spent 40 years as a diligent visual observer of the Moon and planets – who long remained a holdout loath to give up the old methods, a latter-day John Henry dueling with the optical equivalent of the steam hammer – I have at last yielded. As my own eyesight has become less keen, I have caught something of the fascination with CCD. I suppose my eventual capitulation was inevitable – like the use of oxygen by climbers of Everest.[44] My results so far have been modestly successful and encouraging, rather like those of the Sunday painter compared to those of the Old Master. The real Masters of CCD imaging include Don Parker, Ed Grafton, Bill Flanagan's, Isao Miyasaki, Damian Peach, Eric Ng, Tan Wei Leong, Anthony Wesley, men who – rather like rival hunters or Homeric champions clashing beneath the walls of Troy – engage in friendly one-upmanship on the Internet where they routinely post their mind-blowing images. Martin Mobberley, himself a doyen of British imaging, expressed it well at a meeting of the British Astronomical Association: "To be the best in the world, as in any hobby/sport, you literally have to be manically obsessed and with a highly competitive streak."

---

[44] According to Walt Unsworth, *Everest*, p. 78: "George Mallory saw in oxygen a challenge to the human spirit an attack by Science on natural values." Or in the words of George Finch, who advocated its use: "there existed another force of oxygen antagonists, largely unscientific, who were willing enough to admit that oxygen might, indeed, have its uses, but condemned it on the ground that its employment was unsporting and, therefore, un-British."

Most of my own hours at the telescope are now spent with a CCD camera attached where the eyepiece used to go, feeding wires into a computer on the table next to which I sit, using software to stack images, adjust brightness, stretch contrast, apply unsharp masking and employ other techniques of image-processing to bring out the magic of the details. I am not without passion for this work. And yet – not infrequently – I find myself overtaken with the sense that such things are "not mine, and were not made for me."[45] When I feel that strongly, I have no choice but to remove the CCD camera, insert an eyepiece, and look with my own eyeball.

Then the nostalgia for the way things used to be done returns. I recall again how I used to be entranced by the hour with a lovely planet blazing on the night sky through the slit of a small dome: what it was like to commune, one on one, with these other worlds, how I sometimes experienced during those nocturnal vigils the same state of abstraction from the mundane that the adept in Eastern mysticism accomplished only through painstaking mastery of the lotus position combined with rigorous inculcation through all the stages of abstruse contemplation. I know that I have sometimes attained a state approaching Nirvana – one of complete absorption in which I have left behind all my and the world's present troubles and distractions and been wrapped up in my extraterrestrial enterprise. I have then been one with Woodhouse, the astronomer in H.G. Wells's story "In the Avu Observatory" who

> must have forgotten things terrestrial. All his attention was concentrated upon the great blue circle of the telescope field – a circle powdered, so it seemed, with an innumerable multitude of stars, and all luminous against the blackness of its setting. As he watched, he seemed to himself to become incorporeal, as if he too were floating in the ether of space.[46]

Or with Percival Lowell, who wrote of the devotee of the planets:

> Withdrawn from contact with his kind, he is by that much raised above human prejudice and limitation. To sally forth into the untrod wilderness in the cold and dark of a winter's small hours of the morning, with the snow feet deep upon the ground and the frosty stars for mute companionship, is almost to forget one's self a man, for the solemn awe of one's surroundings.[47]

Or with H. P. Lovecraft's Azaroth, from whose casement could be seen

> only walls and windows except sometimes when one leaned far out and peered at the small stars that passed. And because mere walls and windows must soon drive to madness a man who dreams and reads much, the dweller in that room used nightly to lean out and peer aloft to glimpse some fragment of things beyond the waking world and the grayness of tall cities.[48]

[45] Wordsworth, *Prelude*, I, 23.

[46] H. G. Wells, "In the Avu Observatory." In: *Best Science Fiction Stories of H. G. Wells* (New York: Dover Publications, 1966), p. 276.

[47] Percival Lowell, *Mars and Its Canals* (New York: Macmillan, 1906), p. 8.

[48] To be perfectly correct, it was the character in Lovecraft's abortive novel, *Azaroth*, these words describe. The 500 words of the beginning of this novel are quoted in their entirety in a letter from H. P. Lovecraft to Frank Belknap Long, June 9, 1922.

I assert, however, that it is not even necessary to use a telescope for all the old passion to kindle into flame. On winter evenings, as I am writing this (February 2009), Venus looms high in the western sky, and I follow its motion nightly with my naked-eye. I know that the planet is in fact an inhospitable blast-furnace of a world, seared by a runaway greenhouse effect; more *inferno* than *paradiso*. For all that, when I see it I am unaccountably stirred with some of the same nostalgia I felt for it as a child of nine when I wished with all my might on it as the Evening Star and it was like a spirit from fairyland. For me, Venus is not a world of fact as much as a state of mind – it thrills me as it ever did and casts its shadow not only from above but from within.

Chapter 3
Nomads

Their wandring course now high, now low, then hid,
Progressive, retrograde, or standing still.

Milton, Paradise Lost, VIII, 126–127

No one can say just when human beings first noted five bright "stars" moving among the other stars, the *planets*, so-called from the Greek word for wanderers. They have been known from time immemorial, though it was only in Greco-Roman times that they received their familiar names: Mercury, Venus, Mars, Jupiter, and Saturn.

Homage to Scriven Bolton. Venus appears resplendent in the evening sky in the Boundary Waters of Northern Minnesota. Painting by Julian Baum. © Julian Baum

Usually the planets travel in an orderly manner, progressing from west to east; but sometimes they stop, reverse direction, and trace a grand loop-the-loop among the stars.

W. Sheehan, *A Passion for the Planets: Envisioning Other Worlds, From the Pleistocene to the Age of the Telescope*, DOI 10.1007/978-1-4419-5971-3_3,

What were these moving lights? What purpose did they serve?

Most astronomy books begin – at the earliest – with a few words about the Egyptians and the Babylonians before passing on to the Ionian Greeks, who are credited with having been the first to aspire to a rational understanding of the world (thus Thales of Miletus, the earliest Ionian philosopher of whom we have record and one of the Seven Wise Men of Greece, said, "All is water" – however, he also added, "and gods"). But this book is about the *passion* for planets – the thrill some of us have in pursuing knowledge about them, the sources of those motivations and compulsions that can lead to expenditure of vast effort and unreasonable resources.

The passion is as intense as it is because ultimately it is derived from more basic drives serving essential biological needs. Those drives were selected for (though Darwinian natural selection) because they have been indispensable to survival and reproductive success.

Until the last few thousand years, all humans were nomads – like the planets themselves, they did not stay put anywhere for long. Instead they lived in small, mobile groups that followed a "wandering course" across the African savannas or the vast Eurasian landscape that R. Dale Guthrie calls the "Mammoth Steppe." They moved with the Great Beasts, which they hunted with "exquisite pleasure and excitement."[1]

One has a glimpse of the relatively unchanging way of life followed by nomadic tribes for hundreds of thousands of years from ethnographic studies of groups that still live by the Old Ways.

One such group, the Nyae Nyae !Kung – or Ju/wasi as they call themselves, a term meaning "real people" or "first people" – are nomadic hunter-gatherers who live in the Kalahari desert in northeastern Namibia. What is remarkable about them is how little concerned they are with astronomical phenomena. They do not worship the Moon or stars, or pray to them for rain or success in the hunt. According to Lorna J. Marshall, "They regard [all these] as distant 'things of the sky,' beyond man's knowledge."[2] To the !Kung, the Sun represents searing heat, thirst, hunger, exhaustion. It is not beneficent, but regarded as bad – a thing of death. Theirs is the perspective of desert-dwellers. For them, it is just as well that the Sun is far away. If it were any closer, people would burn to cinders. It is rain, not the Sun, that brings life to the Earth.

The !Kung, despite their remarkable perceptiveness and knowledge of the natural world, are said not to have noticed the solstices because of the flatness of their land and the lack of permanent roof-peaks or kraal-posts to fix a spot on the horizon by which to note them. They return after the round of the seasons to the same places but always choose a fresh location on which to build hearths and set up grass shelters. Not even a baobab tree near a water hole is seen consistently from the same

---

[1]R. Dale Guthrie, *The Nature of Paleolithic Art*. Chicago and London: University of Chicago Press, 2005, p. 226.

[2]Lorna J. Marshall, *Nyae Nyae !Kung Beliefs and Rites*. Cambridge, MA: Peabody Museum of Archaelogy and Ethnology, Harvard University, 1999, p. 251.

position relative to the horizon so as to provide a fixed reference point from which to map the Sun's apparent movements.[3] The !Kung also regard the Moon as a "useless thing." In general, they are anti-astrological: they do not believe that celestial bodies influence human affairs. The celestial bodies occupy a realm of their own utterly apart from the earth and separate from and indifferent to man. It seems the !Kung repay that indifference in kind.

Africa has been the cradle of all hominids, including *Homo sapiens*, the species to which we belong and which appeared only 150,000 or 200,000 years ago. Small groups of hunter-nomads, gracile, light-bodied people, once lived all over that continent, their ways of life, well adapted to the relatively unchanging environment in which they found themselves, remaining – as among the !Kung – much as they had always been, in some cases until well into the twentieth century. In the early 1950s, Elizabeth Marshall Thomas (Lorna's daughter) moved to the Kalahari with her parents and spent her formative years among the Bushmen, a small widely dispersed population of perhaps 10,000 individuals living in small groups who subsisted by hunting and gathering in an area of dry bush land and desert perhaps a hundred thousand square miles in extent (thus the population density was about one person for every ten square miles). She recalls the life she led among them:

> I saw the Old Way, the way of life that shaped us, a way of life that is now gone. I also feel that I saw the most successful culture that our kind has ever known, if a lifestyle can be called a culture and if stability and longevity are measures, a culture governed by sun and rain, heat and cold, wind and wildfires, plant and animal populations.... Aspects of this culture were known to the very first members of our lineage, whose bones were found near Port Elizabeth, South Africa, in the Klasies River Mouth Caves, where they had rested for 150,000 years, some of the earliest remains of *Homo sapiens* yet discovered.[4]

Some of the descendents of these "first people" have experienced relatively little change in thousands of generations. Finding themselves well-adapted to their conditions of life, they have continued to follow the Old Way. Indeed, humans are a conservative species by and large; experiments are risky, and except under the pressure of the direst conditions, our tendency is to adhere to traditional ways, in the assurance that worked in the past is likely to work best now and in the future.

We know from ethnographic studies that humans lived in small and scattered groups, their size limited, as they still are in harsh places like the Kalahari, mainly by the availability of sources of water. Thomas notes that in the 6,000 square miles known as Nyae Nyae in northeastern Namibia, there were only seven waterholes considered permanent because they had not failed in recent memory even during drought.[5] The nomads of the Kalahari nevertheless manage to get by with very little water, getting much of what they need from plants and roots.

---

[3] Among the Bushmen, only the Nharo, of Botswana, seem to have recognized the solstices.

[4] Elizabeth Marshall Thomas, *The Old Way: a story of the first people*. New York: Farrar, Straus and Giroux, 2006, pp. 6–7.

[5] Thomas, *The Old Way*, p. 22.

Apart from the need to maintain proximity to water holes, another thing that long limited the range of the small groups of humans on the African savannas was the need not to spread themselves so thin in hunting and gathering that they couldn't make it back to the watering hole and the safety of the camp by nightfall. The day belongs to people, the night to lions and other dangerous predators; the fear of the dark is deep-embedded in the human psyche for good reason. With the exception of humans, notes Guthrie, "most carnivores hunt mainly at night, when they can avoid overheating and when the prey's vision is at a disadvantage. Carnivores work every edge, challenged as they are by the acute senses and physical abilities of their prey species."[6] Humans have poor night vision – a legacy of our primate past in the trees. On the other hand, color vision – which was selected for because of the advantages it conferred on primates as also for tree lizards, birds, and arboreal rodents to locate and determine the ripeness of fruit – provides better discriminatory resolution for seeing prey in the daytime than does black-and-white. (I have used this advantage in some of my planetary observations with larger apertures, 1 meter and more; because of the greater image brightness, details are detected that are invisible in smaller telescopes. The superiority of the large aperture is not its ability to resolve fine details but to bring out subtle nuances of colors.)

On the Kalahari, Thomas experienced first hand the savage loneliness and sheer terror of being caught far from camp as the night begins to come on:

> I am alone an hour's walk from camp, sitting in the long grass at the edge of an arm of the pan, listening to the wind moving the grass and to something going huff, huff, miles away – a lion – and looking at the hazy gray sky. It will be dark soon, and I'm looking at the miles and miles of yellow, silver grass and black bushes in the grass, and thinking how the wind may have blown for thousands of miles before it touched a person, and perhaps it blows over a Bushman camp tucked away somewhere, one point in the enormous, vast veldt that goes hundreds of miles in every direction and it is all like this, just grass and grass and grass, and a few bushes and a few thorn trees and a few antelope in small herds and a few groups of lions and a few groups of Ju/wasi as far apart from each other as the stars – all living in this country but so small and few that they are hardly aware of each other. The wind stops. The air seems very still. The sun is moving down and the sky in the west is yellow. A cold night is coming. I hope I can remember how I came here so that I can find our camp before the night sets in. In all that space, you could miss it by the slightest turn or step, and walk right by.[7]

That, then, is the world that shaped our kind for thousands of generations. Far removed as we may feel ourselves from the Old Times and the Old Ways, they haunt our dreams, and are reflected in deep feelings and preferences we cannot explain. They make up what we call instincts or drives. They are our tropisms and predilections, our innate fears and aversions. They are often mysterious even to ourselves but no less part of our repertoire of responses to the world around us than for other animals. Such are love of water, the comfort of sitting in a circle around the warmth of the fire, *the instinctive fear of the dark.*

---

[6]Guthrie, *The Nature of Paleolithic Art*, p. 218.

[7]Thomas, *The Old Way*, p. 22.

Fear of the dark is among the earliest and most universal fears to appear in childhood. It first emerges at about 2 years of age – about the same time children also have their greatest fear of animals – and it continues to grow in intensity until about five.[8] The fact that for thousands of generations of our species there were very real dangers from the things that went bump in the night makes it rather unlikely that our ancestors went out into the dark alone, unless absolutely necessary, or spent much time serenely stargazing or pondering the movements of the planets. Instead, they must have been extremely reluctant trespassers into the nighttime world – the world that belonged to other predators, whose roars, yelps, and whoops must have curdled their blood.

Even today, most astronomers know the occasional attack of night terrors. I think of E. E. Barnard, alone on Mt. Wilson during the winter of 1905 with the Bruce telescope, taking his wonderful wide-field photographs of the Milky Way. Barnard recalled:

> I must confess that at times, especially in the winter months, the loneliness of the night became oppressive, and the dead silence, broken only by the ghastly cry of some stray owl winging its way over the canyon, produced an uncanny terror in me, and I could not avoid the dread feeling that I might be prey at any moment to a roving mountain lion.... So lonely was I at first that when I entered the Bruce house and shoved the roof back I locked the door and did not open it again until I was forced to go out.[9]

But this deep-seated association of the night with risk and danger may also help to account for the "thrill of the night sky" some of us experience, an emotion for which psychologist William E. Kelly has coined the term *noctcaelador*.[10] And what group of humans, on the savannas in Africa or later on the Mammoth Steppe of Eurasia during the Pleistocene, would have been most likely to experience this emotion most strongly? In all cultures, including our own, the group for which risk-taking activities reach their peak are adolescents, especially adolescent males. Perhaps that is one reason why, for so many, an interest in astronomy – if it is going to take at all – does so in late pre-puberty or adolescence. In any case our motivations for studying the night-sky – and our affinity to the nomads that wander there – are deep. The passion for planets is founded in complex emotions, and "nurtured alike by beauty and by fear."[11]

---

[8] According to C.W. Valentine, summarized in: Jeffrey A. Gray, *The Psychology of Fear and Stress*. London: Weidenfeld and Nicholson, 1971, pp. 16–18.

[9] E.E. Barnard, unpublished manuscript; quoted in William Sheehan, *The Immortal Fire Within: the life and work of Edward Emerson Barnard*. Cambridge: Cambridge University Press, 1995, pp. 340–341.

[10] Kelly has coined this word to describe the very strong emotion of some of adoration and attachment to the night sky. He indicates that individuals who experience it tend to have traits of novelty-seeking, sensation-seeking, openness to experience, and enjoyment of effortful cognitive activity, which may indicate a preference for cognitive complexity. See: William E. Kelly, "Getting a Thrill from the Night-sky: the relationship between sensation seeking and Notcaelador," *Psychology Journal*, 4, 1 (2007), 40–46.

[11] A slight misremembering of Wordsworth here; *Prelude*, I, 302.

For that matter, we modern humans are still wired up neurologically and biologically to the life ways of the savannas of Africa and – for those of European ancestry – to those of the Eurasian Mammoth Steppe of the Ice Ages. The past indeed is prologue. We now find ourselves bound for the stars in large part because our ape-like ancestors swung through the trees and later some of them left the forests for the savannas and still later some of them left the savannas for the Mammoth Steppe. Those great migrations, which led to changed conditions of life and to adaptations still bred into the marrow of our bones, were dictated by severe necessity. That necessity – and the main driver of evolution – was climate change.

The first of a series of significant events that changed the climate of the Northern Hemisphere occurred 2.5 million years ago, when the Himalayas, which had first begun to form 60 million years ago when the Indian Plate slammed into Southeast Asia, had reached so high into the atmosphere that they began to block the monsoonal circulation of moisture from the oceans in the south. As a result, arid conditions set in across a wide swath extending from Europe through Siberia and across the Bering Land Bridge into Alaska and the Yukon – the vast area we have been referring to as the Mammoth Steppe. Then, about 1.6 million years ago, the Epoch of Ice Ages – the Pleistocene – began. The Gulf Stream, which now lends some of the warmth of the Gulf of Mexico to Europe, turned south and smacked up against the coast of Africa. Eurasia cooled, and episodes of glacial advance – at least 20 since the beginning of the Pleistocene – began to regularly bear down on the landscape. Indeed, the last glacial maximum occurred only 18,000 years ago. (Though the last 10,000 years have seen a relatively warm interglacial period, the presence of massive continental ice sheets on Greenland and Antarctica along with numerous smaller glaciers in mountainous regions throughout the world, retreating with alarming rapidity owing to human-induced global warming, shows that the last Ice Age is not quite over even yet.)

The mechanisms behind the sudden climate transitions such as those associated with Ice Ages and episodes of interglacial warming are complex and the details are still being debated. Clearly, however, they are partly tied in with grand cosmic events. Owing to the gravitational pull of the other planets, both the shape (eccentricity) of the Earth's orbit and the tilt of its axis undergo quasi-periodic variations that cause the Arctic region of the Earth (the region lying between latitude 60 and 70° north, near the Arctic Circle) to receive unusually low amounts of summer radiation at intervals of about 90,000 years. This is the so-called Milankovitch Period, named for the Serbian civil engineer and mathematician Milutin Milankovitch who proposed such theory to account for the Ice Ages in the 1920s.[12] This astronomically-driven cooling is amplified by various meteorological mechanisms – for instance, a change in the circulation of the North Atlantic has long been suspected to play a major role. The point is that as the growing glaciers captured more and more of the

[12]The theory was not really verified, however, until deep-ocean cores became available. See: J.D. Hays, J. Imbrie, and N.J. Shackleton, "Variations in the Earth's Orbit: Pacemaker of the Ice Ages," *Science*, 194 (1976), 1121–1132.

Earth's water, aridity increased on the vast Eurasian Steppe. Across this vast, cold, arid steppe-like habitat,

> the climate was too cold, windy, and dry for most trees, but arid- and cold-adapted forbs and grasses could survive. During glacial episodes, as aridity intensified, the Mammoth Steppe spread northward as well as east and west, making most of unglaciated Eurasia an arid grassland, characterized by wind, low rainfall, and cloudless skies that were brilliant in the summer and blue frigid in the winter.[13]

As the moisture was locked up in the growing glaciers, rainfall also decreased in sub-Saharan Africa. Tropical rainforests disappeared, to be replaced in turn by woodlands, savannas, and finally deserts.

From an analysis of mitochondrial DNA of indigenous populations scattered across sub-Saharan Africa, it appears that, following their origin in East Africa 150,000 or 200,000 years ago, humans split into two main groups who lived in isolation from one another for nearly a hundred thousand years; one headed south, the other northeast. A glacial episode beginning about 115,000 years ago and reaching its cold and arid maximum 70,000 years ago seems to have played a particularly decisive role in shaping the destiny of our species by producing a catastrophic drought in sub-Saharan Africa during which humans narrowly escaped extinction. (From DNA studies, it is estimated that the population of *Homo sapiens* world-wide dropped to a dangerously low level – a scant 2,000 individuals!)

Parenthetically, I need to point out that at that time ours was not the only hominid species extant. In Europe, there were the Neanderthals; in the Far East the Ngandong hominids of Java (apparently late survivors of the *Homo erectus* line) and the recently discovered and much debated *Homo florasiensis* found on the Indonesian island of Flores. The latter seem to have become extinct only 13,000 years ago, so that, as Ian Tattersall and Jeffrey Schwartz point out, "As a model to explain the emergence of human beings and their remarkable attributes, the Great Chain of Being [leading from the lowest forms of life to man at the top of the Chain beneath the Angels and God] has once again let us down. There was nothing inevitable about how we got where we are today."[14]

Nothing inevitable. There was that tight evolutionary bottleneck after all. (If we had not made it, perhaps Neanderthals would today be applauding their survival as the only hominid species and considering themselves the apple of the Creator's eye.) Somehow a handful of humans survived, and formed a single pan-African population which then began to fan out again. Some of the surviving bands – perhaps by threading their way along a series of seasonal lakes and surviving on shellfish – managed to migrate out of Africa altogether by means of the Levant and the Middle East.

---

[13]Guthrie, *The Nature of Paleolithic Art*, p. 18. Clearly, the Mammoth Steppe was a frigid desert – rather like Mars – a desert of extreme dryness and cloudless skies. It must have been more forlorn and monotonous than the bleak scene described in the verse from Wordsworth's *Excursion* – "… moon and stars/Glance rapidly along the clouded heaven/when winds are blowing strong" – since there were rarely clouds for the moon and stars to glide among.

[14]Ian Tattersall and Jeffrey Schwartz, *Extinct Humans*. New York: Westview Press, 2000, p. 247.

From thence some moved farther east into Southeast Asia and Australia. Others turned west into Europe.[15] It appears that this epic migration out of Africa occurred between 60,000 and 40,000 years ago.[16]

Consider the assets and liabilities these expatriated Africans carried with them into the new conditions of the Ice Age world. They had, of course, all those interconnected traits which had distinguished hominids from the other apes ever since they first left the forests (where the other apes remained) for more open woodlands and finally for scrub savannas, in response to earlier episodes of climate change, millions of years ago. These traits included

> extensive intellectual capacities, dependence on learned behavior, opportunistic versatility, capacity for complex language and communication, dexterous and versatile hands, material-manufacturing potential, and cooperative social abilities, all of which in degree distinguish us from other apes.[17]

As hunters, they also carried with them that versatile and quintessential weapon, the sharpened wooden spear, probably wielded by their ancestors for at least half a million years, possibly more. Lacking horns or tusks or large canine teeth, man found the spear a great equalizer, and a ready defense against formidable predators such as lions and hyenas. It had sufficient killing power to double as an offensive weapon, and was versatile, able to be thrust as a pike (keeping the action well out front of the body and its vulnerable organs) or thrown as a javelin. In the latter case, the speed of the throw – and hence both the range and the force – can be markedly enhanced by the use of spear-throwers (atlatls).

The ability to wield such weapons was based, in part, on the long previously evolved characteristics of bipedal posture and the firm attachment of the arm at the shoulder with 180° of global motion, a legacy of our primate evolution in which our

---

[15]Perhaps their emigration was aided by the appearance of a favorably mutated gene for the human dopamine D4 receptor linked to novelty-seeking and risk-taking (also, alas, in modern settings, to Attention Deficit Hyperactivity Disorder and addictions) that increased the odds of their embarking on this great and dangerous migratory adventure. An analysis of genetic variations suggests that this mutation occurred recently – within the last 40,000 years; under the challenging conditions of life in the Ice Age, the mutation apparently conferred a significant advantage since its prevalence increased. See Yuan-Chun Ding, Han-Chang Chi, Deborah L. Grady, Atsuyuki Morishima, Judith R. Kidd, Kenneth K. Kidd, Pamela Floodman, M. Anne Spence, Sabrina Schuck, James M. Swanson, Ya-Ping Zhang, and Robert K. Moyzis, Evidence of positive selection acting at the human dopamine receptor D4 gene locus. *Proceedings of the National Academy of Science*, vol. 99, no. 1 (January 8, 2002), 309–314.

[16]According to mitochondrial DNA dating. Russell Thomson et al., "Recent common ancestry of human Y chromosomes: evidence from DNA sequence data," *Proceedings of the National Academy of Sciences* (U.S.) 97;13:7360–7365 (20 June 2000). These authors state that their results indicate that "movement out of Africa occurred around 47,000 years ago. The age of mutation 2, at around 40,000 years ago, represents an estimate of the time of the beginning of global expansion." This was during the depths of the Ice Age, when continental glaciers were advancing across Europe and severely affecting climate not only in higher latitudes but in the tropics as well. The tropical deep Atlantic Ocean cooled 4°C on average during the last glacial maximum, and the cold ocean currents drew moisture-laden air off the African continent.

[17]Guthrie, *The Nature of Paleolithic Art*, p. 221.

ancestors brachiated in the trees. In turn, the male hunting specialization has been invoked to explain why men consistently outperform women in gross-motor, visuo-spatial and similar tasks that support "targeting" activities, i.e., the throwing or intercepting of projectiles. (I suppose that explains why young males spend so much of their leisure time throwing, kicking, swinging at, or catching balls, and why they tend to idolize the paid professionals who excel at such things). It also explains why boys and men tend as a rule to have rather better intuitions about physics than girls and women – though needless to say, girls and women have many cognitive superiorities over men (the ability to read non-verbal social cues and infer states of mind, to mention one or two).[18]

The main targets of these hunters were large herbivorous mammals. In the African savannas that run from Ethiopia to the Cape and are centered on the Great Rift Valley of East Africa, these included great herds of eland and zebra and gazelle. The predators were lion, leopard, hyena, jackal, and man. On the Mammoth Steppe, the corresponding herbivores were the cold-adapted saiga antelope, reindeer, wooly horse, steppe bison, wooly rhino, and wooly mammoth. The predators were wolf, lion, hyena, bear – and once again man.

Humans, with their taller, more gracile African bodies, were adapted to the warm climate conditions of equatorial Africa rather than to the cold as the stocky Neanderthals were. They never penetrated far into the cold, windy treeless plains farther north; instead they remained on the steppe's more hospitable southern fringes. They seem to have found windbreaks and firewood in prairies along edges of woodlands, and sheltered among the limestone hills scattered across large parts of southern Europe. No doubt, like the peoples of the Far North today, they also fashioned tents, and made garments of skin and fur that would have been essential to their survival under such harsh conditions. (Bone needles used for such work survive from as far back as 26,000 years ago, though the fabrics fashioned by means of them are long gone.) The groups were small, consisting of 25–40 people, mostly close kin. They were kept small by high mortality and low fertility rates.[19] Apart from the marked separation of labor between males and females, specialists must have been unheard of. Within their respective sex-roles, each person must have been a generalist. The males hunted, and were proficient in making their own spears, spear-holders, spear-straighteners; women took the lead in child rearing, cooking, sewing and keeping

---

[18] See, for instance, D. Kimura, *Sex and Cognition*. Cambridge, MA: MIT Press, 1999, and S. Baron-Cohen, *The Essential Difference: the truth about the male and female brain*. New York: Basic Books, 2003.

[19] A recent model with possible relevance to the conditions of life during the Pleistocene is that of the Inuit, among whom the food supply was rarely dependable enough to allow them to settle in one place for long and whose population density was very low (approximately 0.03 persons/kilometer in Canada and only slightly higher in Alaska). Factors helping to maintain stability of population size included predation, starvation, disease, accidents, and social mortality (including suicides and murders). Among Inuit males, the major cause of natural death was accidents – hunting accidents among men accounting for 15% of deaths of a southern Baffin Island group. See: William B. Kemp, "The Flow of Energy in a Hunting Society." *Scientific American*, 225, 3 (1971), 104–115.

Pleistocene encampment. Tents pitched on the Mammoth Steppe, beneath the stars of the Great Bear. Painting by Julian Baum. © Julian Baum

up the camp and hearth. So far it is not difficult to picture them or their way of life; the Siberian hunter/nomads and the Eskimo still live this way, or at least did so until quite recent times.

I picture those late Pleistocene men, women, and children, their tents pitched across the vast, cold, arid steppe, under brilliant cloudless skies of summer and blue but frigid skies in the winter. Some of the closer stars – like Sirius and Procyon – because of their proper motions were not then in the same positions they occupy today; but the Pole Star – since the Precessional Cycle is 26,000 years – would have been Polaris when those bone needles were being used, and the constellations that circumscribed it would have been the same, with slight distortions, as ours.

The stars must have been magnificently cold and brilliant; the Moon and the brighter planets moved – like Great Beasts – among the stellar herds, as their reflections trembled upon the Ice. Even on moonless nights, all was not utterly dark; the light we receive from the distant stellar universe is greater than one imagines. They could dimly see, even by weak primate night-vision, by illumination of star and planet and the false dawn of the Zodiacal Light.

What would those wondering and fearing ancestors have made of that spectacle? That we can never really know; like their music and their laughter, the cosmic surmises of the men and women of that time have not survived. Such thoughts are perishable, and survive only when embodied in less vulnerable stuff than fatty visco-elastic substance of the brain. What we can say is that – though they left impressions of the Great Beasts of the Steppe on the walls of the limestone caves they entered – they did not generally depict images of the Sun and Moon and stars (there are only a few possible exceptions to be considered later). Perhaps I am wrong; but I cannot believe that upon those rolling, treeless landscapes, and under such ravishing skies, they beheld the spectacle playing over their heads with complete indifference.

They must have sometimes noticed that there were Nomads and Migrating Herds above them as well as around them. Perhaps – as they did with the caves – they sometimes entered the Night-time world for the sheer pleasure of exploring its vast recesses; only the sky was too far above them to absorb the parietal images they painted in ochre and other pigments, or to take on the impress of the handprints they left as registers of their encounters. They retreated without leaving a trace. But one suspects they must sometimes have seen images of Animals and Hunters as the heat from their fires radiated out toward those twinkling stars and into eternity; they must have told stories, created legends, about them, just as all men, everywhere, do.

They might sometimes have felt some of the sentiments expressed by Lord Byron in his fragment "Darkness," which conjures up not the Beginning but the End of the World:

> I had a dream, which was not all a dream.
> The bright sun was extinguished, and the stars
> Did wander darkling in the eternal space,
> Rayless, and pathless, and the icy earth
> Swung blind and blackening in the moonless air;
> Morn came and went – and came, and brought no day,
> And men forgot their passions in the dread
> Of this their desolation; and all hearts
> Were chilled into a selfish prayer for light:
> And they did live by watchfires….

But unlike those miserable souls in that Ice-Age World at the end of times, the people of the late-Pleistocene knew that their Night would not go on forever. The Sun would rise, the Day – the Hunt – would begin again. Humans would set forth into the world that is their world, the world of Day; then – and for a long time afterwards – they would remain only occasional interlopers, intruders, into the realm of Night, above-ground or below, in sky or cave, that darkling realm ruled by predators and ghosts and the planetary and stellar watchfires of distant, inscrutably wheeling worlds and suns.

Though somewhat cooler on average compared to conditions across Eurasia today, the main difference in the late Pleistocene climate was not absolute cold but rather the extreme instability and unpredictability of the climate. Under these conditions, more than ever before and perhaps more than ever since, human survival depended critically on living by wits. Those groups with the most flexibility, creativity, and willingness to take risks were more likely to be favored by Natural Selection than the rigid traditionalists, tied as they were to stereotyped, hidebound solutions.

At this point it is perhaps necessary to assure the reader that we are getting back to planets, through a "wandering course now high, now low, then hid." I admit that I have taken what must seem like a very long digression. But the digression, I suggest, is not so much off-theme as might appear. I invoke the cycles of the Earth's axial tilt and orbital eccentricity, climate change and Natural Selection and (in admittedly broad strokes) am getting us from the small-brained ape-like bipeds that hunted the savannas of East Africa at the beginning of the Pleistocene to the masterful, creative and fully human intelligences reflected in the brilliant art of the caves of northern Spain and southern France that appeared, in a sudden and miraculous outburst for which we seem to have been utterly unprepared, at the end of it.

The art of the late Pleistocene – the Upper Paleolithic – is certainly among the human wonders of the world. Until the mid-nineteenth century, when the discovery of the Neanderthals scandalized pious Victorian Judaeo-Christian Europeans, the great antiquity of man was about as unsuspected as the true extent of the Galaxy. When one referred to "ancients," one meant the classical Greeks and Romans. That there were much more ancient people – Stone Age people – who were rather more than savage primitives and capable of creating works which, for want of a better term, can be considered "art" was an idea almost impossible to grasp. When Paleolithic art was first revealed by Sanz de Sautola at Altamira, in northern Spain, in 1875, it was "too splendid to be believed."[20] This enthusiast who first attempted to familiarize the world with his findings was greeted with skepticism and ridicule; he died prematurely, a sad and broken man, suspected of being a simpleton or a fraud. He went too much against the grain. At the time, there were still overwhelmingly strong predispositions to believe that modern humans were the exclusive possessors of all the accoutrements of civilization, including art and religion, and it was arrogantly assumed that the aptitudes for them among earlier men would, as in savages, have been present only in rudimentary and childish forms. Only the exceptionally intuitive, like Sanz de Sautola, recognized the art at Altamira was not a forgery. It was decidedly not "modern"; it conveyed a completely different sensibility. Picasso, upon seeing the paintings of Altamira years, said: "None of us could ever paint like that!"[21] Altamira has been dated to 14,480 years from one engraved shoulder-blade (the two earlier phases of ceiling-decoration are believed to be older than this, while the black figures and polychromes are younger). The best-known Paleolithic cave art, at Lascaux, is rather earlier. However, there are now hundreds of Paleolithic sites known, scattered across southern Europe; new ones are still being discovered.

[20]Paul G. Bahn, *Journey Through the Ice Age*. Berkeley: University of California Press, 1997, p. 17.

[21]Quoted in David Lewis-Williams, *The Mind in the Cave: Consciousness and the origin of art*. London: Thames & Hudson, 2002, p. 31.

For instance, Chauvet cave, located at the foot of the Lower Cretaceous cliffs of the Cirque d'Estre, at an ancient meander of the Ardéche river near where it enters the Rhone in southern France, was discovered as recently as 1994. The paintings at Chauvet are among the most spectacular of all, despite being the earliest so far known – they have been carbon-dated to 33,000 years before the present. That is so old as to make one's hair stand on end. When he first saw pictures of these works, the noted art historian Ernst Gombrich could only exclaim: "Man is a great miracle." The art of Chauvet is not artistically inferior to that at Lascaux or Altamira from almost 20,000 years later. Though there are differences – including in the mammals represented – they are clearly part of the same tradition; a tradition that disappears completely after the warming period of the Holocene begins.

At yet another Paleolithic site, Cosquer, someone entered the cave 20,000 years ago and vandalized paintings that were then already 10,000 years old. In turn, the entrance of the cave was sealed off by the rising level of the Mediterranean sea when the Holocene began, some 10,500 years ago.

Chauvet's words describing what he felt on first descending into the mysterious chamber of horse's heads and then into the even deeper chamber of lion's heads are worth remembering:

> Alone in that vastness, lit by the feeble beam of our lamps, we were seized by a strange feeling. Everything was so beautiful, so fresh, almost too much so. Time was abolished, as if the tens of thousands of years that separated us from the producers of these paintings no longer existed. It seemed as if they had just created these masterpieces. Suddenly we felt like intruders. Deeply impressed, we were weighed down by the feeling that we were not alone; the artists' souls and spirits surrounded us. We thought we could feel their presence; we were disturbing them.[22]

The prevailing interpretations of the art of the Paleolithic have long been based on what might be magico-religious interpretations. These paintings on the cave walls were supposed to have been produced by Ice Age shamans, perhaps in the course of leading initiation rites of young males (it is universally accepted that most of the cave art – which is thematically dominated by images of large mammals, such as bison, horses, wooly mammoths, lions, and aurochs – was produced by males). A good deal of this is ingenious; it all seems to hold together, and if one is in the right frame of mind, it seems completely convincing.

I vividly remember sitting in the Faculty Club at the University of Witwatersand, Johannesburg, South Africa, in 2001, and being utterly enthralled by the first-hand account of David Lewis-Williams, Professor Emeritus of the Rock Art Research Institute, of what one experiences in squeezing through the narrow passage leading into one of the caves of Southern France and then coming into the presence of these images of the Great Beasts. He described illusion of movement – of these Beasts coming alive – when seen by the flickering light of a torch fire. I still recall his exclamation: "It blows your mind." Perhaps the high carbon-dioxide concentration

---

[22] J-M. Chauvet, E.B. Deschamps, and C. Hilaire, *Dawn of Art: The Chauvet Cave*. London: Thames & Hudson. 1996, pp. 41–42.

in these caves, where limestone is being dissolved by percolating water, also colludes in producing an altered state of consciousness.

When Chauvet saw the images on the walls of the cave now named for him, he sensed that "time was abolished"; he and his friends could sense "the souls and spirits" of the artists surrounding them. The hitherto hidden images invoked in them less a sense of "scientific wonder" than of awe; even, perhaps, "spiritual" proclivities (e.g., evidence of early shamanism). Such feelings invite the assumption of what Lewis-Williams calls "a congruence in cognitions – between them and us." The sense of congruence is certainly strong, but as L. P. Hartley famously began his novel *The Go-between*: "The past is a foreign place: they do things differently there." How much more must this apply to an era as remote as the late Pleistocene.

The cave art certainly seems full of spiritual significations. Were these the Pleistocene equivalents of the Sistine Chapel? At Lascaux, where there seems to be evidence of the existence of a scaffold to support the painters, the main chamber has long been referred to as the "Sanctuary." If only the walls could speak. But perhaps they do.

All the same we need to remember the tendency of people to read pictures in different ways – true of the art of the Paleolithic, true equally, we shall find, of the interpretation of planetary images. Images remain semantically equivocal; just because we can construct elaborate and self-consistent stories about them – whether magico-religious interpretations of parietal art made tens of thousands of years ago or theories of canal-systems on Mars – we must remind ourselves that it is the nature of human beings to create "Just-so stories" about all manner of things. We have a marked tendency to see faces in clouds. That tendency may partly account for our ability to conjure up gods all around us,[23] and more than likely gave the first impetus to representational art. It is as Picasso once observed:

> If it occurred to man to create his own images, it's because he discovered them all around him, almost formed, already within his grasp. He saw them in a bone, in the irregular surfaces of cavern walls, in a piece of wood. One form might suggest a woman, another a bison.[24]

What I'm getting at is that we mustn't overdo the spiritual motivations here. It is salutary to turn from Lewis-Williams's fascinating *Mind in the Cave* to Dale Guthrie's equally fascinating *The Nature of Paleolithic Art*, with the truth probably lying somewhere in between. Instead of being an art historian, Guthrie has spent 40 years in Alaska where he is professor emeritus at the Institute of Arctic Biology at the University of Alaska in Fairbanks, studying the remains of long-buried wooly mammoths, and he also regularly bow hunts large mammals on the Alaska tundra. Guthrie places the cave art not in the context of shamanism but in the evolutionary niche of the energy-scarce Ice-Age world. He argues that those who created the images of the Great Beasts of the Ice-Age were hunters – probably mostly young

---

[23] Stewart Guthrie, *Faces in the Clouds: A new theory of religion*. New York: Oxford University Press, 1993.

[24] Pablo Picasso; quoted in R. Dale Guthrie, *The Nature of Paleolithic Art*, p. 134.

males, whose sense of pride, social stature, access to women, and quantity of well-cared for children would all have been determined by their success in the physically violent activity of hunting. They did not hunt in anger – hunters rarely do; nor is there any evidence that they warred with other groups, as post-Neolithic tribes did, nor that they spent much of their time going about dazed with religious fantasies. Hunting, to be successful, is hard-headed, profoundly empirical, and imaginative, rather like science itself. Those who were involved in it no doubt found excitement in the risks and dangers of hunting the Great Beasts. They must have been keen observers of game, pursued with the intensity of obsession to a degree which we today would find difficult to imagine. The most accomplished of them would have succeeded in becoming, in some sense, almost one with the beasts they hunted.

They also seem to have had a penchant for drawing. Perhaps, Guthrie suggests, drawing was common to the individuals in those late-Pleistocene groups as writing is to us today. They drew what interested them the most – the Beasts which were the objects of their preoccupation. Such was the ecological niche they occupied that their survival depended on a "close approximation between the outer reality and our inner perception and interpretation of that reality."[25] Such a close approximation is attested in these closely observed and often vividly realized images which suggests that those who made them were not in a trance but very wide awake.

It is important to emphasize that the two views are not mutually exclusive. Though hunting is a highly practical endeavor, it probably did occur in the context of those spiritual proclivities we, who are still wired the same way, seem to sense so strongly in the cave art. Such proclivities obviously go back a long time – we are hardwired for it as for creativity, art, the passionate pursuit of game, tracking, targeting, throwing projectiles. All of that was forged, along with so much else, in the crucible of the late Pleistocene. We have both the hunt, and the spirit of the hunt.

Certainly the marvelous efflorescence of Paleolithic art – which is uniformly representational art – must be related to the dramatic extension of cognitive abilities that provided selective advantages during the wildly fluctuating and challenging conditions of the Ice-Age world when there were "repeated abrupt swings from scarcity to richness and diversity and density of large-mammal species."[26] Humans had long exhibited strong aptitudes for play, imagination, virtual-reality testing, flexibility. The unusual alignment of circumstances in the late Pleistocene pushed things further in that direction; the riot of creativity reflected in the art of the Upper Paleolithic introduced a new frontier of human adaptation in which imagination was critical. The capacity to produce this art – which seemed to burst forth with such suddenness as to seem astounding and inexplicable – reflected not some cluster of symbols having secondary meanings but rather a sharp intensification in the neurocircuitry of human imagination giving those who had it a selective advantage over those who didn't. Among the byproducts of that intensified imagination were art and religion. The imagination was the ground of all that.

---

[25] Guthrie, *The Nature of Paleolithic Art*, p. 12.

[26] ibid., p. 397.

So now, having circled around the question and now back to it again, we ask what, if anything, these more imaginative humans might have made of the great pageant playing overhead of Sun, and Moon, and planets? Did they pay any attention at all; or were those bodies the things that were in the category of "noted as present but irrelevant"?

Among the Nyae Nyae !Kung, as we have seen, celestial phenomena were in that latter category. Like the people of the late Pleistocene, they did not remain fixed in place for long, and would not have been in a position therefore to make the careful measurements needed to establish the kind of astronomical knowledge of later, more settled peoples. However, there is at least fragmentary evidence of incipient astronomical interests even in the Paleolithic.

According to the somewhat controversial interpretations of Alexander Marshack, marks on bone and antler – once fancifully regarded as "hunting tallies" – are interpreted as notations of recurrent astronomical phenomena such as the phases of Moon which, he argues, would have been the principal means available to late-Pleistocene people for measuring the passage of time. Also, according to Paul G. Bahn:

> An ivory plaquette from Mal'ta (Siberia) has hundreds of pits engraved on it in spirals; these have long been interpreted as symbols of moon worship. Frolov claims that there are seven spirals, the biggest of which has 243 pits in seven turns, and the others 122 (243 + 122 = 365). By analogy with similar motifs found among modern Siberians, he interprets it as calendrical ornamentation (243 days being the gestation period of the reindeer, the staple food at Mal'ta, as well as the length of the winter), while the summer lasts about 122 days.[27]

In addition to keeping track of the lunations, they must have noticed the movements of the brighter planets. Given their aptitude for interactive perceptions – their uncanny ability to take some bump or crevice of a rock-face as suggestive of the curved contour of the form of an animal, then finish it with a few strategic strokes – it would be hard to believe that they never saw figures outlined among the stars.

Were they mere outlines, or did they represent spiritual forces? Was their sky a thing of naught, as for the Nyae Nyae !Kung, or of intense interest and significations – "full of animals and spirit guides," as the German archaeologist Michael Rappenglueck has suggested? According to the anthropologist Evelyne Lot-Falk:

> Hunting is not a luxury activity but a problem of vital importance, requiring an attention, concentration and seriousness which the modern huntsman necessarily lacks. It is not a duel between man and beast in which man succeeds in reassuring himself of his enormous technical superiority. It is not merely a confrontation of adversaries: behind the animal the hunter sees a host of supernatural forces ready to intervene against this irruption into their domain. The roles are reversed: man is puny and seems to be attacking a strong opponent

[27] Paul G. Bahn and Jean Vertut, *Journey Through the Ice Age*. Berkeley: University of California Press, 1997, p. 205. The reference is to: Frolov, B.A. Numbers in Paleolithic graphic art and the initial stages in the development of mathematics. *Soviet Anthropology and Archaeology*, 16, 1977/1978, 142–166.

> from a position of inferiority. He is faced with a warrior whose resources are no less than his own, who is under the protection of supernatural powers.[28]

Only by invoking the same supernatural powers that protect animals could Lilliputian man subdue these Brobdinagian beasts. Technique alone could accomplish nothing; it was efficient only when backed by appropriate rites – in other words, by magic. Where did they look for such help? Most often, no doubt, they looked within their own breasts; but it is also likely that they sometimes turned for help to the mysterious sky.

I would like to imagine that they dreamed onto the Cave of the Night a set of Upper Paleolithic constellations of the same Great Beasts of the Ice Age that they depicted in the subterranean walls of their limestone caves. There is at least some evidence, faint but suggestive, that they did.

They would certainly have been more sensitive to nuances in the colors of the stars than we are. The color red would have been particularly portentous for hunters. In addition to the representational art of the caves, rows of red-ocher dots are ubiquitous on their walls, and their significance has long been one of the most baffling and seemingly insoluble problems of Upper Paleolithic art. Are they some kind of code? Guthrie, with his usual astuteness and inspired by his direct experience of hunting, sees them as blood spots shed by animals not quite fatally wounded which can form a trail leading the careful tracker to the point where the animal finally falls.

The late-Pleistocene trackers of blood trails must have noted the marked ruddiness of Mars, of Aldebaran, of Antares. What did they make of these red dots against the sky? Was Mars a gore-bespotted tracker of celestial Great Beasts? Were the stars themselves blood tracks shed along trails across the Mammoth Steppe of the sky? Is Guthrie perhaps right in thinking that they were a kind of "running documentary"?

> A hunter relies on tracks to determine which species were in the vicinity, how long ago, what direction they were moving, how many of each kind, and their age and sex. A skilled tracker uses tracks to actually picture the moving animal – not the dot-dot-dot of tracks but the animal itself. One can see its speed, tension, and gait … Tracks, like a name or insignia, are an inseparable part of the animals' identity and form.[29]

Were the planets they tracked the hunters, or were they the beasts?

One can imagine Mercury as a swift-footed chamois or ibex, Venus as a white-flanked aurochs, Mars as a blood-splattered lion, Jupiter as a bear, Saturn as a wooly mammoth. But we will never know by what names the hunters of the late Pleistocene called them. One can also imagine Great Beasts outlined in the star-patterns of the night sky. Here we are slightly luckier; one of our star-groups – Ursa Major, the Great Bear – may well be one of theirs.

---

[28]Quoted in Mario Ruspoli, *The Cave of Lascaux: the final photographs*. New York: Harry N. Abrams, 1987, pp. 63–64.

[29]Guthrie, *The Nature of Paleolithic Art*, p. 268.

Mars and the Pleiades conspire above the Great Dome of the Lick 36-in. refractor. Photograph by William Sheehan, October 2005

According to many versions of the myth, the Bear consists of the four bowl stars, followed by the three stars in the handle representing hunters. The same star/myth combination is shared not only by the Greeks, Basques, Hebrews and Siberian tribes in the Old World but by the Cherokee, Algonquin, Zuni, Tlingit and Iroquois in the New. According to Bradley E. Schaefer, "the Bear is unlikely to be an independent invention, because the stars do not look like a bear."[30] Not, I would add, unless one has bears on the mind. For tens of thousands of years the hunters of the late Pleistocene contended for the caves with a particularly ferocious rival species, the Great Cave Bear (*Ursus spelaeus*), which reached lengths of 20 ft and had huge teeth and razor-sharp claws. The cave bears heavily clawed the protrusions and alcoves of these caves; for instance, at Chauvet the claw-marks of a bear have been incorporated into the so-called Panel of Horses. Bear skulls and bones are found in abundance, and in one recess – known as the Skull Chamber – a bear skull seems to have been purposefully placed on the flat surface of a block fallen from the ceiling.[31] The bears themselves are among the animals represented on the walls. Though long-extinct on Earth, one of the Great Cave Bears may yet roam the skies – remembered

[30] Bradley E. Schaefer, "The Origin of the Greek constellations," *Scientific American*, November 2006, 96–101:98. I'm reminded of what Winston Churchill said on a visit to Lick Observatory: "We all know how astronomers have mapped the heavens out in the shape of animals. We can most of us – by a stretch of the imagination – recognize the Great Bear, but still one quite sympathizes with those who call it The Plough. Bear or Plough – one is like it as the other."

[31] Jean Clottes, *Chauvet Cave: the art of earliest times*. Salt Lake City: University of Utah Press, 2003, p. 100.

in a starry petroglyph. If so, it may well be one of the oldest extant inventions of humanity.

Another possible Paleolithic star-group is Taurus, the Bull – or rather Taurus, the Aurochs (a now-extinct species of bull). The so-called Aurochs no. 18 at Lascaux appears to have a small group of dots over its shoulder in the same positions as the stars of the Pleiades. Here and there one finds another set of spots suggesting a possible star-group; in a cave in Spain there is a handprint with a circlet suggestive of Corona Borealis, the Northern Crown. More fanciful is the idea that the Bull, Birdman, and Bird on a Stick, forming perhaps the most dramatic panel at Lascaux, represent the stars of the Summer Triangle which never set from the Northern Skies some 17,000 years ago. Here, more than elsewhere, we seem to see what we look for.

Lascaux: Hyades and Pleiades appear above the shoulder of one of the Great Aurochs. Painting based on the original Upper Paleolithic artwork by Randall Rosenfeld. © Randall Rosenfeld

Recalling again my own intense feelings for the night sky, the eagerness with which I tracked and hunted the Great Beasts that moved across it, I realize that I, even as one in city pent, experienced there and there alone something of the exhilaration of the Mammoth Steppe. I never hunted, never experienced the excitement of an actual kill. I never had any desire to shed blood. But I experienced intense excitement in seeing a planet I had not seen before, the thrill of a beautiful conjunction of the Moon with a planet in the same part of the sky. I realize there is a deep connection between my passion for the planets and theirs for the large-mammal hunting which forms the central unifying feature of the art of the Upper Paleolithic. The beasts in each case are different; the man is the same.

They covered their caves with images of wooly mammoths, wooly rhinos, lions, bears, deer. I covered the walls of my bedroom with spacecraft pictures of the Moon, Mars, Jupiter, and Saturn. We have here the same underlying drive but it is – through the processes of displacement or sublimation – capable now of being directed to different objects. The Ice-Age hunter – the artist of the Upper Paleolithic – still lives in me; but I have found surrogate Beasts to hunt. The sense of thrill, of risk and danger, of adventure and exploration, of wanting to capture a huge trophy and cover myself in glory, is still there; it has been for me from at least the age of nine. But instead of pointing spears, I have become adept at pointing telescopes. Projectiles, now as then, have been a preoccupation; the atlatl of the present-day is the spacecraft-tipped rocket. When I was 12 or so, I fired water-propelled rockets with the same ardor with which I peered at the planets in my telescope, and dreamed of going into space myself by some such means.

The game is afoot; the hunt is on – pursued with single-mindedness of purpose as ever – with concentration on detail to the point of obsession and with the utmost attention to the techniques and tools needed for success. I chase the greatest and most glorious beasts imaginable. In the pursuit of those beasts – the planets – it may well be that our very survival will depend.

As we can have no real knowledge – only guesses – of the Paleolithic constellations, we can only sketch the religion of those long-ago days. It would be tempting to imagine that, as profoundly involved with the natural world as they were, they had some of the profound ecological intelligence of the Oglala holy man, Black Elk, who in telling his story said it was not his own life that was important but what he had shared with all life: "It is the story of all life that is holy and it is good to tell, and of us two-leggeds sharing in it with the four-leggeds and the wings of the air and of all green things."[32]

Probably they did involve themselves in some form of shamanism. Shamanism – the word comes from the Tunguso-Manchurian šaman, "he who knows" – is based in part on the idea that certain individuals, medicine man, priest, or psychopomp, can enter trancelike states during which his soul or spirit, taking animal-form, travels to other worlds in order to bring the souls of the sick back or escort the souls of the dead to their new abodes in the world or worlds beyond. Perhaps the origin of this is ultimately the intense identification of master-hunters with the animals they seek – a process by which they seem to become one with them. In any case, in order to bring the souls of the sick back, or to escort the souls of the dead to their new abodes in the world beyond, the shaman undertakes daring journeys, full of dangers and trials. Though most closely identified with the Central Asian and Siberian regions where it has its most complete manifestations, some form of shamanism seems to be universal among hunter-nomadic peoples worldwide. There is no reason to doubt that such ideas already existed during the late-Pleistocene, though perhaps the shamanistic elements in Cave Art have been somewhat exaggerated. (The presence of

[32]Quoted in Wendell Berry, "Think Little," in: *A Continuous Harmony: essays cultural and agricultural*. Washington, DC: Shoemaker and Hoard, 1972, p. 82.

anthromorphs – figures with the heads of animals and the bodies of men – in the cave art, such as the sorcerer of Le Trois Frères, may well indicate shamans; on the other hand, there are alternative – more prosaic – explanations. Might they not, for instance, be images of hunters wearing skins in disguise to get closer to their elusive prey?)

In shamanism, we have constellated ideas thousands if not tens of thousands of years old. They still retain something of their old power. For instance, they awakened and stirred the scientific passion of the Chief Designer of the Russian Space Program, Sergei Korolev. As his mother, Maria Nikolaevna, recalled:

> He liked my stories.... We "flew" together on a fairy-tale magic carpet and saw Korek-Gorbunek (the humpback pony), the grey wolf, and many other miracles beneath us. This was enthralling and wonderful. My son would press close to me, would look wide-eyed into the sky, while the silvery moon peeked out from amidst the small clouds.[33]

The pioneering anthropologist Franz Boas recounts stories told by the Eskimo shamans, whose ecstatic capacities allowed them to undertake journeys "in spirit" to any part of the cosmos, flying like birds, spreading their arms before them as a bird does its wings. Boas describes one Baffinland Eskimo shaman who was carried to the moon by his guiding spirit (a bear). He took the precaution of having himself bound with ropes in order to assure his return from the sky. Eventually he came to a house with a "strait gate" (consisting of a walrus jaw) remaining open only an instant; if he were caught in its toils, he would be rent to pieces. But he managed to squeeze through. On entering the house, he found the Man in the Moon with his wife, the Sun. After many other such adventures, he returned to the earth. While he had been experiencing his ecstatic journey, his body had remained lifeless; only now, with the return of his spirit, did it begin to stir once more. At last he unbound himself and recounted the incidents of his voyage to his audience. Boas notes that such exploits were undertaken by the Eskimo shamans for no apparent motive; but they felt the need for them "because it is above all during trance that he becomes truly himself; the mystical experience is necessary to him as a constituent of his true personality."[34]

So did the Baffinland Eskimo shaman really fly beyond the Moon?

Ethnographer Mark Watts has reminded me of a conversation with the Yaqui shaman reported by Carlos Castaneda. After experiencing a vision flight upon taking jimson weed, Carlos asked the shaman, "Did I really fly?" The shaman looked at him in disbelief. "What do you mean? Of course you flew!" Carlos, still not satisfied, asked again: "If I had put a ball and chain on my leg, would it have made any difference?" "Of course," replied the shaman, "You would have had to carry the ball and chain."[35]

---

[33]Quoted in James Harford, *Korolev: how one man masterminded the Soviet drive to beat America to the Moon*. New York: John Wiley & Sons, 1997, p. 17.

[34]Franz Boas, "The Central Eskimo," Washington, DC: *Bureau of American Ethology, Annual Reports of the Smithsonian Institution*, 1888.

[35]Mark Watts to William Sheehan; personal communication, December 4, 2006.

That reminds me also of a conversation I had with Story Musgrave, the veteran of six Space Shuttle launches. I had asked him whether he had trouble sleeping the night before a launch. He replied, yes; he was very nervous and worried. He knew how dangerous it was. But despite the danger, he had to go. "It was the only way I had of getting up there."

One can say the same of the shamans. Their ecstasies – adventures of soul-travel outside the body – were the only way they had, 10,000 years ago, of getting themselves into these Other Worlds.

Mircea Eliade, whose 1951 book is still a useful anthology though it has been criticized for painting with an overly broad brush and for basing its conclusions on excerpts of others' writings rather than on actual fieldwork, recounts a number of aspects of Shaman cosmography.[36] According to Eliade, the Turko-Tatars of the Volga River area of Central Asia envisaged the sky as a tent; they saw the Milky Way as the "seam" of the tent, the stars as "holes" for light. The Yakut of northeastern Siberia similarly saw the stars as "windows … openings provided for ventilating the various celestial spheres."

The sky was also sometimes seen as a lid, not fitted perfectly to the edges of the earth and allowing the great winds to blow in through the crack. Through this narrow crack heroes and other privileged beings, such as shamans, can squeeze their way into the sky.

A tent pole or pillar holds up the center of the celestial tent (and defines the Axis of the World). Here we have a notion that may well reach back to the Upper Paleolithic that is reflected in a biblical passage written perhaps 10,000 years later – "he hath stretched out the heavens like a curtain" (Psalm 104:2). (Galileo tried to argue in his "Letter to the Grand Duchess Christina" that this particular passage was not meant to be taken literally!)

Several Siberian tribes and also the Lapps, Finns, and Estonians called the Pole Star the "Sky Nail." But the most suggestive image of all is that of the Buryat, the northernmost of the Mongol nomadic tribes, who pictured the stars as a herd of horses. For them, the Pole Star – the "Pillar of the World" – is the stake to which the starry horses are tethered by invisible or magical ties. Inevitably, this image recalls the horse panel at Lascaux.

Of course, the Pole Star changes over time, because of the precession of the equinoxes, a phenomenon first discovered by the Greek astronomer Hipparcos in the second century BC. Some 13,000 years ago, the brilliant white star Vega was the "North Star"; at the time the Great Pyramid of Giza was built, the axis was pointing toward Thuban, in Draco; now, of course, it is the second magnitude star Polaris.

The only star that would seem to be really meritorious of being called the "Nail Star" or "Pillar of the World" is Vega, which at magnitude 0.0 is one of the brightest stars in the sky. It may well be that the idea of a "Nail Star" or "Sky Nail" contains a dim recollection of the end of the Pleistocene 13,000 years ago, when the glaciers,

[36] Mircea Eliade, *Shamanism: archaic techniques of ecstasy*. Princeton, NJ: Princeton University Press, 1964.

which had reached their last maximum 18,000 years ago, were beginning to retreat and Vega reigned as the Star of the North.[37] Perhaps then the "opening," or "hole," through which the World Axis held up the Tent of the Sky was Vega. Through such a hole, shamans believed, the gods descended to earth and the dead to the subterranean regions. (It is interesting, by the way, that many legends describe the portal to the heavens as the North polar region; also, the familiar spirits of many of the Arctic shamans are bears.) By means of this hole the soul of the shaman flies up or down in the course of his celestial or infernal journeys. He experiences a sense of descending into a vortex or of ascending in flight such as may intrude upon the mind's empire during the period of benign confusion – of hypnagogic hallucinations, free-associations, fantasies – that is the entry-point in the dream-world each night. This altered state of consciousness marks a transition where the categories of conventional, logical thought, ordinary space and linear time are suspended, and what Joseph Campbell calls the "weird adventure" can be experienced. The shaman, with his dancing and his drum, is able to will his ecstatic flights into this familiar and yet disorienting realm.

The sky as a tent, the underworld or the heaven as realms entered through a hole, are the concepts that define the basic cosmology of hunter-nomad tribes. But they also correspond with the basic concept of the world with which every child begins life, for even the most sophisticated cosmology of modern science is a modified version of the basic schema of the tripartite cosmos which Mircea Eliade takes to be the basic cosmology of shamanism but whose universality suggests it must be even deeper – and hardwired into the human nervous system. As Paul Fussell writes, this tripartite vision "is so ancient in … myth, religion, and folklore that there is no tracking it to its origins. By the time of written documents it already has an infinite history."[38] Of course, eventually the concept becomes so elaborated and so overwritten that it may not even be recognized as the deepest level of a much-worked over palimpsest.

This is the point. The human mind itself is in every way an archaeological site; therefore, the most in-depth explorers of the past must make themselves, in some sense, cognitive archaeologists, spelunkers of the mind, engaged in the bold attempt to delve not only into artifacts and potsherds but into systems of meaning and the beliefs that informed them. Our modern astronomy – the sensibility that informs it,

[37]Ethnographer Mark Watts has made a close study of the "rainbow arch" panel at Utah's Rochester Creek. He writes (personal communication, December 4, 2006): "The panel I have been focusing on for the past year is one of several that I'll feature in *Journey of the Hero Twins*, a book on Native American genesis myths and associated sky-lore…. This detail shows the Twins' father sitting just below the arch, who, according to the myth, is Sky…. In the schematic view of the heavens revealed by the myth (and the prior work of Jon Polansky) this figure occupies the position of Vega, suggesting the myth contains some memory of the time when Vega was 'father of the sky.' In about 13,000 BC Vega spun around the celestial pole 'held' by one of the smaller stars in Lyra, a 'star-on-a-string.'"

[38]Paul Fussell, *The Great War in Modern Memory*. London: Oxford University Press, 1975, p. 127.

the drive and the passion that created it – is founded on complexly stratified structures of the human mind. We can trace the works of modern astronomers back to the works of the Ionian Greeks, the Babylonians and Assyrians, the Egyptians, the Sumerians; we can excavate deeper foundations – remnants of past cognitive structures, habitations of thought, now jumbled together, haphazardly mingled and confused – reaching from the dimly-remembered time when the Nail Star was Vega and the starry Great Bear first shed its fearsome shadow across the floes of ice. Joseph Campbell says of civilization and so we may say of science which is its peculiar (and fragile) manifestation: "the roots … are deep, our cities do not rest, like stones, upon the surface."[39]

In giving short shrift to the long preamble of the prehistoric past in order to get to the seemingly meatier substance of, say, Greek astronomy or the Copernican revolution, we risk sharing the fate of the German romantic Heinrich Schliemann who – baffled by the complex stratifications of the mound of Hissarlik in Anatolia but inflamed with his boyhood passion to discover Troy, the city of Homer, of which he had so long dreamed – unawares cut through the very layers he had sought. We can excuse him on the grounds that "nothing so complex had ever been excavated, and Schliemann had to learn his technique as he went along."[40] In the same way, though astronomy proper may begin with the Sumerians or the Egyptians or the Babylonians or the Greeks, the passion for it began much earlier. Its flames were no doubt kindled among those who watched by the hearth fires, hunted on the grasslands, and wandered under the cold blue skies of the Mammoth Steppe, and wondered at the planets shining there.

[39] Joseph Campbell, *Primitive Mythology: the Masks of God*. New York: Penguin, 1998, p. 6.

[40] Michael Wood, *In Search of the Trojan War*. New York: Facts-on-File, 1985, p. 57.

Chapter 4
Innana's Antics

Lo, all our pomp of yesterday
Is one with Nineveh and Tyre!

Rudyard Kipling, *Recessional*

The Holocene warming – which fits quite well with predictions based on the Milankovitch cycle – marked the end of the last Ice Age, just about 10,000 years ago.[1] It brought with it momentous changes in plant and animal life together with changes in the ways of humans depended upon them, as well as changes in their relationship with the sky. These changes preceded the Neolithic – or Stone Age – the most important manifestation of which was the agricultural revolution.

Though it is often assumed that the domestication of animals and plants was adopted as a labor-saving innovation, Lorna Marshall and others have pointed out that agriculturists tend to work harder than hunter-nomads. Nevertheless, on the whole agriculture leads to more efficient production of food (it's about ten times more efficient at extracting calories from the soil) and thus can support communities with populations at least ten times higher than those of bands of hunters and nomads.[2] It is also a more stationary way of life. Farmers become wedded to the land they work and need to remain in place through the cycle of planting, growth of plants, and harvest. They are not, like hunter-nomads,

[1]There are several cycles involved. The most important is the obliquity or "tilt" cycle, with a period of about 41,000 years, over the course of which the tilt of the Earth's axis varies between 22.1 and 24.5°. About 10,000 years ago, the Earth was at its maximum tilt of 24.5°, and the Northern Hemisphere experienced maximum heating. Added to this was the fact that 10,000 years ago the Earth was at perihelion during the boreal summer rather than winter. In combination, these effects would have provided about 8% more solar radiation to the Northern Hemisphere in summer.
Admittedly, though the warming trend fits well with prediction, the maximum climate response was delayed – probably owing to feedbacks related to ice from the previous Ice Age whose maximum ended only 18,000 years ago – and the warming period occurred earlier in some locations, especially in the far-southern areas, than in others.

[2]Thom Hartmann, *Attention Deficit Disorder: a different perception*. Grass Valley, CA: Underwood Books, 1997, p. xxvii.

W. Sheehan, *A Passion for the Planets: Envisioning Other Worlds, From the Pleistocene to the Age of the Telescope*, DOI 10.1007/978-1-4419-5971-3_4,

temporary inhabitants of a place, restless and always on the move; they are not eager to pull up stakes to move on, and in place of the mobile circle of the fireside we have first the walled encirclement of a village and next the armed fortress of the city. Instead of remaining ephemeral tenants of a particular piece of ground without notions of personal property like hunter-nomads, agriculturalists take possession of it and develop a sense of conquest, manifest destiny, and ownership (and begin to worship gods who reflect that new way of thinking, who "contract" with their chosen peoples and deal in real estate transactions by offering them "promised lands").

The result of all this is that, ever since the beginning of the agricultural revolution some 9,000 years ago in the ancient Middle East, agriculturalists have pushed into the wilderness wherever they found it, exploiting the land and removing the hunter/gatherer peoples standing in their way, so that today "fewer than 2% of the world's human population are genetically pure hunter/gatherer peoples, and only a remnant of them is found in our gene pool, and that only as the result of enslavement and assimilation."[3]

This process continues apace. It has completely transformed humans in their relations to one another and to the planet they inhabit. "Go forth and multiply," the command in Genesis, is that not of hunter-nomads but of agriculturalists for whom the domination of the Earth and its inhabitants including both animals and other humans becomes not only acceptable but foreordained and sanctioned by the gods (or by the men through whom the gods speak to men). Much of the Old Testament contains gruesome descriptions of bloody conquests. E. O. Wilson, in his book *Creation*, points out:

> Civilization was purchased by the betrayal of Nature. The Neolithic revolution, comprising the invention of agriculture and village, fed on Nature's bounty. The forward leap was a blessing for humanity. Yes, it was: those who have lived among hunter-gatherers will tell you they are not at all to be envied. But the revolution encouraged the false assumption that a tiny selection of domesticated plants and animals can support human expansion indefinitely. The pauperization of Earth's fauna and flora was an acceptable price until recent centuries, when Nature seemed all but infinite, and an enemy to explorers and pioneers. The wildernesses and the aboriginals surviving in them were there to be pushed back and eventually replaced, in the name of progress and in the name of the gods too, lest we forget.[4]

The early Neolithic farmers – those in the so-called pre-pottery Neolithic, whose settlements in the Zagros Mountains of central Iraq date back to about 9000 BC – subsisted on a diet consisting mainly of domesticated sheep, goats, and cattle. Among staple plants were grasses that flourished with the onset of the Holocene – wild and domesticated species of einkorn, emmer wheat, barley. Needless to say, this was a radically different diet from that which nourished the hunter-nomads of the Upper Paleolithic.

---

[3]Ibid., p. xxviii.

[4]E. O. Wilson, *The Creation*, p. 11.

These farmers slashed down trees and laid out rice paddies and wheat fields. As they did so, they changed not only the landscape but the atmosphere itself. Indeed, global warming owing to human activity is nothing new; it seems to date back to the beginnings of agriculture itself. By laying waste to the wilderness, these early farmers caused major alterations to levels of greenhouse gases such as methane and carbon dioxide and probably even forestalled the onset of another ice age. As a result, global temperatures – slowly falling again around 8,000 years ago – thereafter began once more to rise.[5]

The earliest settlements consisted of rectangular houses built of pressured lumps of clay, and for a long time the inhabitants still used stone tools. The religions of such people also continued at first to be shamanistic. Among probable shaman images of the Neolithic (Stone Age) are plastered skulls found at Jericho, in Anatolia (modern Turkey), where the inhabitants kept goats and cultivated cereals 9,000 years ago. Seashells represent eyes; presumably – as with the eyes of red scoria and obsidian added to the great stone heads of Easter Island for festival days – the eyes were the source of their power, and turned these effigies into "a living face." Since they were buried with the dead, they may well have been meant to emphasize the importance of seeing even in death.

Among the most remarkable of these effigies are the highly stylized statuettes of human figures found not at Jericho but at nearby 'Ain Ghazal (6750 BC), in which the eyes bulge slightly and are surrounded by marked oval ridges of black bitumen. In 2004, when I was at the British Museum in quest of the Babylonian Venus tablet, I chanced across the 'Ain Ghazal figures, and was mesmerized by them. These lifeless things seemed somehow animate, alive.

The figures from 'Ain Ghazal, gazing wide-eyed across the centuries, as if possessed with boundless wisdom, may not so fancifully be regarded as the distant predecessors of the Babylonian priest-astronomers or even of such modern astronomers as Tycho Brahe and William Herschel, who shared the same intense passion for seeing into the beyond. Herschel even said, in the spirit of the shaman: "I have seen farther than any other man." "Their ability to 'see,'" writes archaeologists David Lewis-Williams and David Pearce, "was beyond anything that ordinary human beings could experience … frighteningly omnipercipient."[6] As such, they may be regarded as some of the earliest patron saints of astronomy, which has always been a visual science and has depended on the percipience – if not omnipercipience – of its seers.

Modern-day shamans, such as the Tungus of Siberia and the San of South Africa, enter into relationships with spirit-animals, from which they speak of deriving

[5]Robin McKie, How prehistoric farmers saved us from new Ice Age. *The Observer*, March 6, 2005: "Computer models of climate developed at the University of Wisconsin-Madison indicate that the rise in carbon dioxide and methane would have had a profound effect on Earth's climate; without man's intervention, the planet would be 2°C cooler than it is now, and spreading ice caps and glaciers would affect much of the world."

[6]David Lewis-Williams and David Pearce, *Inside the Neolithic Mind*. London: Thames and Hudson, 2004, p. 75.

'Ain Ghazal shaman. British Museum. Photograph by William Sheehan in 2004

enabling "electricity." Lewis-Williams and Pearce describe the case of a San woman, "a healer and a shaman of the game," who kept her "animal helper," a castrated springbuck (a medium-sized South African antelope) tied up by means of a thong so it could not "wander about." This is suggestive. These authors even suggest that the domestication of animals might have begun not as a means of increasing the production and availability of food but with the more mystical attempt of shamans to control their spirit-animals and to keep them, too, from "wandering about."[7]

One animal that may have been among the first to come under control of the early agriculturalists was the aurochs, the skulls and horns of which are prominently

[7]Ibid., pp. 143 ff.

displayed in the interior rooms of the dwellings of Çatalhöyük, another Neolithic village in Turkey. An impressive painting on the wall of another house at Çatalhöyük shows an enormous red bull surrounded by a group of tiny, insignificant-looking male figures. The relative size of the aurochs and the humans suggests the awe with which the aurochs must have been regarded.[8] Though sheep and goats already seem to have been domesticated at Çatalhöyük, domesticated cattle do not yet appear to have been a prominent part of life at the site despite the prominence of bull imagery – the implication being that at Çatalhöyük we may actually be witnessing an early stage in the long process of domestication.

Mircea Eliade even suggests that the first fortifications around villages – trenches, labyrinths, ramparts – may have been designed as magical defenses against demons, rather than as means of repelling attacks by human beings.[9] Certainly there are symbolic – even magical – aspects to the dwellings at Çatalhöyük, which were constructed to reflect the three-tiered cosmos to which their inhabitants subscribed; beneath the floors they buried their dead, while their houses were entered by means of an aperture in the roof which they could access by means of ladders.[10] This hole also became the outlet for the smoke of their hearths.[11] At 'Ain Ghazal, where the wide-eyed statuettes were found, "people constructed rectangular structures with semicircular apses at one end. The apse seems to have been a focus of interest, because in one instance the center of the arc is marked by an orthostat (large standing stone)... The attention of people entering the apsidal buildings was directed, or controlled: they were led to see what the designers of the structure intended them to see – the curved space of the apse and the adjacent orthostats. Guided (or controlled) sight is a key feature of religious experience."[12]

In such architectural details, we see an early manifestation of the tendency – later, the compulsion – to orient structures to significant points in the sky. It would find its consummation in the megalithic monuments of Stonehenge and the Great Pyramids of Giza. But already, in these early Neolithic structures, the erection of orthostats – heel stones, kraal-posts, gnomons – guide or control the sight to significant celestial positions: the Sun at the summer-solstice, the Moon or Venus at their most

---

[8]Ian Hodder, "Women and Men at Çatalhöyük," *Scientific American*, January 2004, 77–83.

[9]Mircea Eliade, *The Sacred and the Profane: the nature of religion*. Translated by Willard R. Trask. New York: A Harvest Book, Harcourt, Brace & World, 1959, p. 49.

[10]Compare the belief of the Chukchee of northeast Asia. They hold that "there are several worlds situated one above another, in such a manner that the ground of one forms the sky of the one below. The number of these worlds is stated to be five, seven, or nine."

[11]See Eliade, op. cit., pp. 56–57: "The house is not an object, a 'machine to live in'; it is the universe that man constructs for himself by imitating the paradigmatic creation of the gods, the cosmogony." He adds (pp. 57–58) that "the most ancient sanctuaries were hypaethral or built with an aperture in the roof – the 'eye of the dome,' symbolizing break-through from plane to plane, communication with the transcendant."

[12]Lewis-Williams and Pearce, *Inside the Neolithic Mind*, pp. 95–96.

An Ikeya-Seki like comet appears over Stonehenge in early Neolithic Britain. Painting by Julian Baum. © Julian Baum

northerly or southerly risings. In the accretive way of myth – in which layer upon layer of mental associations gather just as in the physical world the strata of one habitation or dwelling comes to overtop another – the aurochs in time would become associated with the Moon (whose horned crescent obviously resembles the aurochs with its gigantic curved horns) and presciently but more inexplicably, Venus.[13]

[13]So did the ancients manage to make out the planet's crescent phase? There are a number of credible – but many more not credible – reports of naked-eye achievement of this difficult feat in modern times. But though the phase can be easily detected with binoculars, personally I can't believe that it is possible with the naked eye, or that this is the most plausible explanation for figures of Venus with horns dating from antiquity. On the whole I am inclined to agree with Lick Observatory astronomer W. W. Campbell that ancient descriptions of Venus as a crescent was a "pure lucky guess, probably made under the influence of the crescent Moon."

If the tethering of spirit-animals may have been the first step toward domestication, so the devising of sight-lines to the stars may be regarded as the first step toward the domestication of the heavens. Henceforth the planets and stars are no longer quite "wild": the strange luminous celestial denizens are no longer without names or connection to human beings, they are no longer, as the Nyae Nyae !Kung regarded them, "distant 'things of the sky,' beyond man's knowledge."

We have indicated, tentatively and speculatively, the possibility that the people of the late-Pleistocene saw animals and spirit-guides in the night-sky. Those efforts to populate the sky with animals may not have been very systematic, and probably – until the Neolithic – the sky remained, no less than the earth beneath it, a wilderness. Only in the Neolithic – in heaven as on earth – is the wilderness tamed and domesticated; the starry beasts are corralled within specific provinces, they are tethered (as if by mystic bonds or magical lines of force) to the Earth.[14] Just as the aurochs become domesticated as cattle, the stars are brought under symbolic human control.

The early stages of this process are perhaps dimly remembered in the very term "Zodiac," a translation from the Greek *zodiakos kyklos* – "circle of animals." It would be attractive to believe that the earliest constellations all represented shamanic animals (guide spirits). The earliest of all, as we have seen, was probably Ursa Major, which may go back 40,000 years. Another early constellation must have been the Bull or the Aurochs. (As we have seen, one of the aurochs at Lascaux is shown with a group of dots almost certainly meant to represent the stars of the Pleiades.)

Of the 12 zodiacal constellations currently recognized, not quite half (5 of 12) have nothing to do with animals: the anthropomorphic figures Gemini, Virgo, Sagittarius, Aquarius; also Libra, the balance beam. All of these non-animal constellations are likely to have been Neolithic in inspiration. Anthropomorphic images (including many popularly described as Mother Goddess figures) are common among the artifacts of the Neolithic, and though human figures, the so-called Venuses, are not unknown among the portable art of the Upper Paleolithic, in the Neolithic they occur in profusion. This increased fondness for representing the human form in art probably reflects the increased population-density of humans in villages. The large number of human figures seen in the bull painting at Çatalhöyük has, for instance, no analog in Upper Paleolithic cave paintings. There is also a change in the way they are represented: the bull is less carefully observed than had been the norm in the Upper Paleolithic – as if the artist were drawing what he "knew" rather than what he "saw." The humans are even more feebly drawn. They are poor spindly things, and assume that clumsy quality that is typical of a

[14]Albert Einstein has noted that as long as human beings were tied to the Earth, our observations could never directly reveal to us the "true" planetary motions, but only the intersections of the lines of sight (earth-planet) with the "fixed-star sphere." These intersections first began to be consistently observed during the Neolithic. Forward to Stillman Drake's translation of *Dialogue Concerning the Two Chief World Systems*. Berkeley, CA: University of California Press, 2nd revised edition, 1967, p. xv.

modern child's efforts – they deserve to be considered the world's first "stick men." What we see is a reshaping of the brain itself in response to the changed conditions of life, a first move in the direction from the representational to the symbolic.

According to the Russian archaeo-astronomer Alex A. Gurshtein, the zodiacal constellations were probably defined against a backdrop of earlier constellations that had already been recognized in the dim distant past. He begins with the basic three-tiered cosmology of shamanism – with Lower, Middle, and Upper Worlds – and argues, plausibly, that the earliest star-groups were defined according to a scheme reflecting the same arrangement. Flying beings – the Dragon, the Flying Horse, the Swan, the Eagle – were projected into the stars near the pole. Anthropomorphic beings and land animals, such as the Hunter and the Twins, were relegated to the middle sky-world. Water characters, such as the River and the Fishes, were submerged among the stars that never rose far above the horizon.[15]

Gurshtein argues that by the sixth millennium BC – somewhat after the first settlements appeared in Anatolia – the starry zone of the ecliptic had been mapped out. The ecliptic marks the *via Solis*, the Sun's annual path among the stars, which is also that traversed by the Moon and planets (in contrast to the stars which have by now been tamed and platted out, the Moon and planets remained – and would long remain – willful, unruly, wild things). Naturally, the key points of the Sun's path were superimposed upon the Upper, Middle and Lower strata as defined above. The Sun, Moon and planets thus remained confined to the realms of anthropomorphic and watery creatures, and never touched the upper stratum reserved for the airborne beings.

The four main points along the Sun's path are as follows:

- *Vernal equinox*, also called – for historic reasons – the first point of Aries, marking the beginning of Northern Hemisphere Spring.
- *Summer solstice*, the beginning of summer.
- *Autumnal equinox*, the beginning of autumn.
- *Winter solstice*, the beginning of winter.

Gurshtein argues that about 5600 BC, when the first great agricultural civilization was becoming established in the valley of the Tigris and Euphrates in Mesopotamia (from the Greek word for "land between the rivers"; now in Iraq), these four points fell among the star-groups of Gemini, Virgo, Sagittarius, and Pisces, respectively.[16] The associations suggested by the Sun's trans-celestial journey in fact gave rise to the creation of those constellations. The Heavenly Twins, the offspring of the Sun-god, were identified with the revival of nature in the spring. The Mother-goddess,

[15]Alex A. Gurshtein, "Prehistoryof Zodiac Dating: three strata of Upper Paleolithic Constellations," *Vistas in Astronomy*, vol. 39 (1995), pp. 347–362.

[16]The term Mesopotamia properly refers to the area of the "fertile crescent" between the Tigris and Euphrates rivers, north and northwest of the bottleneck at Baghdad in modern-day Iraq; by extension it is used to describe the whole region from the Zagros Mountains on the northeast to the Arabian desert on the southwest.

figure with an ear of corn in her hand, was associated with summer fertility and only later, in Greece, became a chaste virgin. The Centaur, a hunter on a horse with a bow in his hands, aimed his arrows at the Sun which, fatally wounded, dropped into the Underworld (the Fishes).[17]

This First Celestial Quartet reflected the stages of the annual solar journey in 5600 BC. Slowly, as the pointing of the Earth's axis changed with precession, the alignment shifted, and new constellations were needed to mark the important points. Three millennia later, when the Great Pyramid of Giza was being built near the Nile delta, the four points had shifted to Taurus, Leo, Scorpio, and Aquarius. At present – for precession has continued its inexorable march – the vernal equinox (still known anachronistically as the First Point of Aries) occurs when the Sun is in Pisces, the summer solstice when the Sun is in Gemini, the autumnal equinox when the Sun is in Leo, and the winter solstice when the Sun is in Sagittarius. In other words, since the beginning of human civilization, the Quartet has advanced only a quarter of the way around the Zodiac. All that time, and we are only through the spring season of the Precessional "Great Year."

So far we have been making wide and rather nomadic sweeps along the vast frontiers of human prehistory, drawing hints from enthnographers, cognitive archaeologists, and the hunters and sifters of dreams.

We fast forward – like shamans flying across time as well as space – to conditions of life several thousand years on from the first settlements at Jericho and 'Ain Ghazal and the Zagros Mountains. Now the great agricultural civilizations of the Middle East – the Egyptian and the Sumer-Akkadian – are coming into the dawning of their full glory.

As always, these developments took place against a backdrop of relentless climate change. The end of the Ice Age had been marked by a change from extreme aridity to exceptional humidity in the Saharan and sub-Saharan belts. In Lower Egypt and the Mediterranean littoral, wheat and barley made their first appearance, having been introduced there from southwest Asia. Over a period of several thousand years (a process not finally completed until the third millennium BC) the Saharan belt had dried out to become the "Great Sand Sea." Henceforth it could no longer support the large and scattered populations such as had occupied – during the same period when Eurasia consisted of the vast and arid Mammoth Steppe – much of northeastern Africa. As the climate continued to change, the local inhabitants began to form relatively large population centers around a series of lakes that formed here; still later, when the lakes dried out in turn, they migrated into the Nile Valley running from the cataracts of Upper Egypt to the delta of Lower Egypt. Here the annual flooding of the Nile produced fertile fields. Indeed, the fertility of the land regularly produced bumper crops of grains (emmer wheat and barley at first). These bumper crops formed the basis of Egypt's great wealth and led eventually to sway over a vast empire. And there was the rub.

[17]Alex A. Gurshtein, "On the origin of the zodiacal constellations," *Vistas in Astronomy*, vol. 36 (1993), pp. 171–190.

Bumper crops produced a marked change in the ways of life – a change that would lead to hierarchical societies, specialized roles, and astronomy. Studies of modern hunter-gatherer societies show that they are remarkably egalitarian; meat is scrupulously shared – not, it should be pointed out, because hunter-nomads are selfless, but because they live in small groups in which the behavior of successful hunters is strongly controlled by the will of the collective.[18] This is the way of life of small groups of people living as they have for thousands of generations under conditions that provide only a bare subsistence. There is no reason to think things were different during the late-Pleistocene. Now, however, we come upon something entirely new in human history: a significant surplus is acquired that – instead of being evenly distributed as before, in which case the bonus per head would be marginal – it becomes the prerogative of a privileged minority. (One of the indicators of this more complex economy is the appearance of storage containers, i.e. pottery. Pottery vessels begin to occur in profusion from the seventh millennium BC; their patterns and motifs, incised or painted, serve as the most reliable means of archaeological dating before the introduction of writing.)

Now, though most of the (able-bodied) members of the community are still working full-time at food production, a few are released from the oppressive burden of such tasks. Instead of each person – at least within the boundaries of their sexually dimorphic roles – having to be a generalist and all-rounder as had been true ever since humans were nomads wandering the African savannas, specialists emerge: rulers, administrators, soldiers, astronomer-priests, artisans. Such a community soon ceases to be very egalitarian; it evolves into a social hierarchy, which is apt to become quite unegalitarian. It also evolves into a system, a "machine." A "system," writes psychologist Simon Baron-Cohen, is "anything which is governed by rules specifying input–operation–output relationships"; it is "finite, deterministic, and lawful." It follows that

> Once you have identified the rules and regularities of the system, then you can predict its workings absolutely. This holds true even for more complex systems, where there are many more parameters, or where the rules are much more elaborate. But the rules are in principle specifiable.[19]

These early civilizations would emphasize organization into hierarchical relationships and expend a great deal of effort into specifying the rules (embodied in texts, many of which turn out to be little more than lists and rule-books). This form of thought has been with us ever since, and defines the pervasive legalistic and bureaucratic aspects which are inseparable from modern life. It seems natural, if sometimes irksome, to us; but it is important to emphasize that it is a way of thought completely alien to hunter/nomads. Their world is – by comparison – unpredictable and subject to wild vicissitudes. They must always be ready to respond to the

---

[18] David Sloan Wilson, *Darwin's Cathedral: Evolution, Religion, and the Nature of Society*. Chicago: University of Chicago Press, 2002, p. 21.

[19] Simon Baron-Cohen, *The Essential Difference*, pp. 61–62.

environment and exhibit responses that cannot be predicted (especially during periods of dramatic climatic fluctuations, which are not predictable). During the more than million years of the Pleistocene, there had been nothing like the perennial inundations of the Nile to provide a backdrop (presumably divinely ordained) of predictability and lawful order.

The perennial drama of the inundations of the Nile led to an emphasis on continuity and tradition expressed in the deep-dyed conservatism of Egyptian culture. The same forms in religion, art, and social organization appear time and time again over the centuries, even the millennia, of that great civilization.

Inevitably, because the reliable and predictable flooding of the Nile followed an annual cycle, there were astronomical tie-ins. The Earth and the heavens now imaged each other, and there was a correspondence in their laws. The Nile flooded when Sirius, the Star of Isis, first appeared in the morning sky before the Sun. The rising of Sirius in August thus became the most important observation in ancient Egypt; the very livelihood of the people was thought to depend upon it. Plutarch, the Greek writer of the first century AD, wrote in his treatise *Isis and Osiris*: "Of the stars, the Egyptians think that Sirius, the Dog Star, is the Star of Isis, because it is the bringer of water." The annual inundation of the Nile was also the basis of the Egyptians' reckoning of the seasons (*akhet* was the season of inundation, *peret* that in which the land emerged from the flood, *shomu* that of the drought). It led early to the adoption of the first solar calendar in history. The lunar calendar is universal among nomadic peoples; however, the Egyptians supplanted it with a calendar consisting of 12 months of 30 days each, with five additional days at the end of the year. The historian of ancient astronomy Otto Neugebauer has referred to this as "the only intelligent calendar which ever existed in human history."[20]

The Egyptian religion was extremely complicated, and involved the worship of many gods. Among the chief gods was the Sun, worshipped as the god Râ. The Egyptians of the earliest era seem to have imagined the universe in the form of a box, the bottom of which was narrow, oblong, and slightly concave (centered, naturally, on Egypt itself). The great river, the Ur-nes, marked the ecliptic, the Sun's path through the heavens, and it flowed through the mountains. In the north it was hidden behind them as it ran into a valley, the Daït, where surrounded in endless night it became the heavenly Nile – the Milky Way.

Along the river floated a boat whose passenger was a disk of fire, the Sun itself (Râ). The same stream carried the bark of the Moon (Iââhu; sometimes called the left eye of Horus) which appeared out of the "door of the east" in the evening. This stream also carried the planets. While Ûapeshetatûi (Jupiter), Kahiri (Saturn) and Sobkû (Mercury) steered their barks in a forward direction, like Râ and Iââhu, Doshiri (Mars) sometimes oared backwards – a striking reference to the retrograde movements the planet exhibits when it is nearly opposite to the Sun in the sky.

[20]Otto Neugebauer. *The Exact Sciences in Antiquity*. New York: Dover, 1969 reprint of 1957 ed., p. 81.

Stairway to Heaven at Kom Ombo, leading to the gallery from which the Egyptian priests studied the stars. The temple at Kom Ombo (known in ancient times as Nubt; the City of Gold) is located on the Nile 50 km north of Aswan (the ancient Syene), and was built by the Ptolemies in the second century BC. Painting by Julian Baum based on a photograph by William Sheehan in 2001. © Julian Baum

But of all the planets, dazzling Venus most captured the imagination of early observers. The Egyptians regarded it as a close confederate of the Sun, lagging behind it in the evening when the Egyptians called it Uati – the first star of the night – and advancing ahead of it in the morning when it was known as Tiû-nûtiri, the harbinger of the Sun. Yet another name for it was Benin, the heron, a bird still common along the banks of the Nile and which dives under the river, only to rise again. In the same way Venus, the celestial heron, disappears into the Underworld sometimes for many months at a time but always returns to take its place as leader of the starry host.

In contrast to the Egyptians, the people – or rather peoples – of Mesopotamia, in which arose the other great agricultural civilization of antiquity, never adopted a solar calendar; they remained strict lunarians. Though in Egypt the annual inundation of the Nile imposed itself on every aspect of human life, in Mespotamia conditions for agriculture were more variable. The annual rainfall is low; the ground becomes dry and hard, unsuitable for the cultivation of crops, for 8 months of the year, while the sluggish flow of water in the rivers deposits large quantities of silt and elevates the river beds to the point where the waters overflow their banks or change their course. Mastery of the situation was eventually only achieved through the creation of an extensive system of artificial canals. Thus, according to Hans J. Nissen and Peter Heine, in contrast to the Nile, which deposits its fertile sediment on the land just before the sowing season, guaranteeing a high yield every year without the need to fertilize artificially or to let the land lie fallow, the Tigris and Euphrates

> overflow only when they pose a great danger to the harvest. While the Nile flows south to north, the twin rivers of Mesopotamia … flow in the opposite direction. Since these rivers run high as a result of melting snow and ice in their catchment basins, the melting phase in Ethiopia [the source of the Blue Nile] occurs much earlier in the year than it does in south-eastern Anatolia, the source of the Euphrates and the Tigris. Still, there can be no doubt that only abundant water in the rivers enabled the intensive agriculture so essential for early civilizations. In Mesopotomia, the positive use of water had to be wrested from nature, a task that presented no small challenge for Mesopotamian society.[21]

Only with the introduction of irrigation and drainage about 4000 BC were the fertile lands between the Tigris and Euphrates able to produce in sufficient abundance to support a large population and then – and then only – were those lands which James Henry Breasted called "the fertile crescent" poised to become a cradle of civilization.[22] Indeed, it has been said that the great achievement of the fourth millennium was the city.[23] The organization of irrigation networks required coordinated effort on a hitherto unattempted scale, and led to vastly larger settlements than the small villages of the early Neolithic. (The very word *civilization* is derived, by the way, from the Greek word *civis*, town dweller.) Again, these developments led to a decisive change in human life. In contrast to developing the global, intuitive, and pictorial processes of the Right Hemisphere of the brain, which had been the forte of the Paleolithic hunter-nomads, the agriculturalists of Neolithic Mesopotamia organized their society ruthlessly according to the logical, analytical, verbal- and rule-based cognitive processes of the Left Hemisphere. Everything was now linearly and hierarchically arranged. In place of the broad egalitarianism of the

---

[21] Hans J. Nissen and Peter Heine, *From Mesopotamia to Iraq: a concise history*. Chicago and London: University of Chicago Press, 2009, pp. 2–3.

[22] A survey of ruined settlements around ancient Hatra, about 300 km north of Hatra, has shown that the southern limits of the zone in which agriculture is possible without artificial irrigation has remained unchanged since the first settlement of Mesopotamia.

[23] Among the Sumerian cities was Ur, the Biblical "Ur of the Chaldees," the birthplace of Abraham, which was excavated by Sir Charles Leonard Woolley; also Kish, Erech, Nippur, Larsa, Eridu, Lagash, Umma, Tello.

past, human life came to be centrally controlled, rigidly organized, systematized. The Mesopotamian city also became the prototype of the same regularizing principle:

> The city, with its plumb lines and right angles, represented the straight lines of logic, not the winding ways of intuitive, emotional thought. The cityscape was ruled in the lines of its building, and it was also ruled in order while the wilderness without was unruly....[24]

Lands were "conquered," animals "domesticated," people defeated in wars and enslaved. Even the gods were effectively "settled" in temples. All was to be ordered and organized.

That, in any case, was the aspiration. I must admit that the whole process – in retrospect – seems to have been rather confusing. The history of Mesopotamia is as long as that of Egypt but maddeningly more involved. One gets a picture of those great empires that is rather grand but gauzy, composed of scenes like those of the huge historical paintings fashionable in the last century showing the rise and fall of ancient empires – but seen by moonlight. Egypt was an empire of the Sun; the successive empires of Mesopotamia wax and wane in succession in emulation of the phases of the Moon. Thoreau strikes a fine pose with the following passage (little known since it is not from *Walden*):

> The light of the moon! In what age of the world does that fall on the earth? It was as the earliest dewy morning light, and the daylight tinge more reminded me of the night. It was such a light, perhaps as sufficed for the earliest ages. Yet a light not used and worn by man like that of the day, but stranger and more affecting. There were the old and new dynasties contrasted, and there was an interval between greater than between the most distant epochs recorded in history, an interval which time cannot span. On that illustrated sandbank was revealed an antiquity beside which Nineveh is young. The Silver Age is not more distant from the Golden Age than moonlight is from sunlight; yet perchance nations have flourished in that light....[25]

Sumerians, Akkadians, and Sumero-Akkadians who founded their Empire around Ur (the "Ur of the Chaldees" that was the birthplace of the biblical Abraham) were followed in turn by Babylonians, Assyrians, Chaldeans, Persians, Greeks, Romans, Parthians, Sassanids. Still later by Muslims, Ottoman Turks, and the British – the latter, in 1920, combining the former Ottoman provinces of Basra, Mosul and Baghdad into "Iraq," a country that had never existed before; finally, in 2003, by Americans and British.

Names bubble up from long-ago readings of general histories or the Bible of kings who were once worshipped as gods but now hardly remembered as men – Sargon, Ashurbanipal, Nebchadnezzar, Sennacherib. Faint memory-traces are retouched and revive from brisk walks through the Assyrian exhibits of the British Museum (en route to see the Elgin Marbles) or pleasant afternoons spent, during graduate school, at the Oriental Institute of the University of Chicago among

[24] Joy Griffiths, *Wild: an elemental journey*. New York: Jeremy P. Tarcher/Penguin, 2006, p. 34. The villains are wild animals and rustic people. Griffiths notes (at p. 35) that the word "villain" (a Middle English variant of *villein*, "peasant") once meant a rustic, and "the root of the word is in villa – originally the word was merely a simple description of where someone dwelled. The word gradually shifted, coming to mean criminal."

[25] Henry David Thoreau, *The Moon*. Boston and New York: Houghton Mifflin Company, 1927, pp. 10–11.

Lamassur that once guarded the throne room of the Assyrian king Sargon II (eighth century BC). Photograph of the sculpture in the Oriental Institute at the University of Chicago by William Sheehan, 2008

winged bulls (lamassurs) and curly-bearded warriors. One manages in this way to conjure up a period of great, if vague, magnificence (but perhaps great magnificence is necessarily bound to be vague; detail can only diminish the effect – as we have seen to some extent even with the planets). The official portraits of the kings or warriors are without clues to their personalities; with nothing to humanize them, after a while one grows weary and moves on to the next exhibit. At the end of the day one has a picture of the inexorable hand of entropy. One sees trembling across the desert air a mirage-like vision of desolate ruins.

Though it has become a cliché that the shores of history are strewn with the wrecks of empire, we should remember that the attempt to impose order on a wilderness was once original; it remains a human aspiration, and began here. The rectilinear plan of streets and canals sprang up on what had once been a mud land. For that matter, the ordering of the chaos was never complete; in contrast to the area around Baghdad, southern Mesopotamia was a region of extensive marshes and reed swamps known the *hawrs*, which defied attempts at reclamation until recent times – Saddam Hussein largely drained them after the 1991 Gulf War, and though partly restored in 2003, they remain under stress because of persistent drought. The American occupiers of Iraq – the latest imperial power to invade the Land Between the Two Rivers – have made their own contribution to the devastation. At Babylon, American and Polish forces built a military depot over the objections of archaeologists. Somebody even tried to gouge out the decorated bricks forming the famous dragons of the Ishtar Gate. After noting the damage done by American helicopters to the fragile palace of Nebuchadnazzar, the Temple of Nabu and that of Ninmah, and other structures, Chalmers Johnson adds:

> … none of this even begins to deal with the massive, ongoing looting of historical sites across Iraq by freelance grave and antiquities robbers, preparing to stock the living rooms of Western collectors. The unceasing chaos and lack of security brought to Iraq in the wake of our invasion have meant that a future, peaceful Iraq may hardly have a patrimony to display. It is no small accomplishment of the Bush administration to have plunged the cradle of the human past into the same sort of chaos and lack of security as the Iraqi present. If amnesia is bliss, then the fate of Iraq's antiquities represents a kind of modern paradise.[26]

Inevitably, the manner in which life came to organized between the Two Rivers led to the development of the world's first bureaucracy. Here indeed was "one of the most tenacious legacies of the ancient East."[27] Like later bureaucracies, the maintenance thereof required the making of extensive rules and the keeping of meticulous records. Need we doubt that the whole process must have made major inroads – annexations, conquests, invasions – into the cerebral cortices of those that maintained them? Language itself was bureaucratized. It is, after all, one of the functions of traditional syntax, as George Steiner has observed, to "organize our perception into linear and monistic patterns."[28]

The records were kept on baked clay tablets. The difficulty and inconvenience of producing curved lines in clay with the slanted edge of a reed stylus used to make the marks in the clay led eventually to the stylized wedge-shaped linear writing characteristic of later cuneiform; perhaps that also contributed to what sometimes comes across as a bland accountant mentality of the Babylonian society. It is at least somewhat disappointing to discover that the earliest literature in the world, developed by the first literate people, does not consist of poetic effusions

[26] Chalmers Johnson, *Nemesis: the last days of the American Republic*. New York: Henry Holt, 2006, pp. 52–53.

[27] History of Mesopotamia and Iraq; *Encyclopaedia Britannica*, 15th edition.

[28] George Steiner, "The Retreat from the Word"; in: *George Steiner: a reader*, p. 297. Elsewhere he points out that "an explicit grammar is an acceptance of order: it is a hierarchization, the more penetrating for being enforced so early in the individual life-span, of the forces and valuations present in the body politic." "Future Literacies," in ibid., p. 431.

or philosophical musings but of long lists of mundane things like heads of cattle or bundles of reeds – though of course much the same would appear of our own society centuries hence if archaeologists unearthed our records offices, banks, government institutions. We are true heirs to our Mesopotamian forbears.

The priests, who created the basis of astronomy, were also bureaucrats of a sort. They climbed their terraced ziggurats (literally, "cosmic mountains") whose seven stories were meant to represent the seven planetary heavens. They assumed their posts on elevated observing platforms which – to carry forward the conceit – were supposed to represent the summit of the universe. (Perhaps it was not so mind-boggling as it sounds; the kings of Akkadia, at least from the time of Sargon I's son's reign, were wont to style themselves "the kings of the universe.") The massive ziggurat of the ancient Sumerian city of Ur, built in the period 2112–2095 BC and restored by Nebuchadnezzar II in the sixth century, is the most famous. (Sadly, during the U.S. occupation of 2003, it was spray-painted by U.S. marines with their motto *Semper Fi* – from *semper fidelis*, "always faithful" – then placed "off limits" in order to hide the desecration that had occurred there. For some reason, but mainly no doubt the complete ignorance among the military planners of the precious legacy of this and other ancient sites, the U.S. military chose the sensitive land immediately adjacent to the ziggurat to build its huge Tallil Air Base with two runways; "in the process," writes Chalmers Johnson, "military engineers moved more than 9,500 truckloads of dirt in order to build 350,000 square feet of hangars and other facilities for aircraft and Predator

The famous ziggurat at Ur in Mesopotamia. In the backdrop shines the Star of Marduk, the chief Babylonian god (Jupiter). Painting by Julian Baum. © Julian Baum

unmanned drones. They completely ruined the area, the literal heartland of human civilization...")[29]

Did the priest-astronomers experience wonder – awe – elation? Did they feel a passion for the planets as they monitored them through nightly vigils? Was their mentality that of enthusiasts or of accountants and lighthouse keepers?

I have seen some of the baked clay tablets that record astronomical data, and read the books written by those who have acquired the skill to decipher them (well aware of the fact that writing itself was invented by the ancient Mesopotamians). I appreciate their importance yet I cannot find much sense of enjoyment in them. The Babylonians were the first Baconians; they practiced the inductive method, collected data and found patterns therein which would later reveal all manner of good things. But for all that, the Babylonian priest-astronomers must have been rather drudge-like; dull obsessives, engaged in the same kind of mind-numbing routine as that with which others diligently kept tabs on commercial transactions. They observed the comings and the goings of the great gods of the universe – divulged their commentaries on the affairs of the king and state – as if they were so much trade in cattle or bricks. (That, alas, was not only true 3,000 years ago. During the eighteenth and nineteenth centuries, the assistant astronomers at government-sponsored observatories such as Greenwich or Paris were also drudges, engaged in routine observation of positions or the reduction of piles of observations with tables of logarithms, and died young, broken by their labors.)

It has been said that "from its beginnings in Sumer before the middle of the third millennium BC, Mesopotamian science was characterized by endless, meticulous enumeration and ordering into columns and series, with the ultimate ideal of including all things in the world but without the wish or ability to synthesize and reduce the material to a system. Not a single general law has been found and only rarely the use of analogy."[30]

---

[29]Chalmers Johnson, *Nemesis: the last days of the American Republic* (New York: Henry Holt, 2006), p. 51. What happened there and elsewhere – for instance, Babylon – during the American occupation makes extremely depressing reading. According to Johnson, pp. 51–52, though U.S. military ruined Ur for further archaeological research, "they did, however, erect their own American imperial ziggurats. On October 24, 2003, according to the Global Security Organization, the army and air force 'opened its second Burger King at Tallil. The new facility, co-located with [a] Pizza Hut, provides another Burger King restaurant so that more service men and women serving in Iraq can, if only for a moment, forget about the task at hand in the desert and get a whiff of that familiar scent that takes them back home.'" The American record elsewhere in Iraq was equally execrable. Thus at Babylon, says Johnson, "American and Polish forces built a military depot, despite objections from archaeologists. John Curtis, the British Museum's authority on Iraq's many archaeological sites, reported that, on a visit in December 2004, he saw 'cracks and gaps where somebody had tried to gouge out the decorated bricks forming the famous dragons of the Ishtar Gate' and a '2600-year-old brick pavement crushed by military vehicle's.'" The one part of what George Bush and Tony Blair had called, on April 8, 2003 "the patrimony of the people of Iraq," protected by the oil-obsessed Bush Administration was the country's oil fields and the Oil Ministry in Baghdad. One can only say, Alas!

[30]*History of Mesopotamia and Iraq*; op. cit.

It cannot be denied that they achieved considerable skill in mathematics of the limited kind accountants use. Their mathematics was based on the sexagesimal system – which explains why even today we have 12 constellations of the Zodiac and why we divide time and angular measurements into hours, minutes, and seconds. They progressed thus far but no farther. They were limited ultimately by the workings of a bureaucratic mind. Neugebauer sums up the extent of their mathematical achievement:

> In spite of the numerical and algebraic skill and in spite of the abstract interest which is conspicuous in so many examples, the contents of Babylonian mathematics remained profoundly elementary... Babylonian mathematics never transgressed the threshold of prescientific thought.[31]

Within these rather severe limitations, there is no denying that Greek astronomy was built on Mesopotamian records, which preserved an enormous amount of raw data. This data included highly exact periods of the Moon and the planets based on careful observations over many centuries.

Whereas, as we have seen, the hunters and nomads of the late-Pleistocene identified at most a few constellations, and even the Greeks as late as Homer's time (eighth century BC) were familiar only with a few of the most obvious asterisms (thus Homer and Hesiod mention only Orion, Ursa Major, and the Hyades and the Pleiades), the more star-conscious civilizations of Egypt and Mesopotamia systematically set out to organize the entire starry wilderness into constellations just as they had tried to organize their alluvial floodplains into cities. The Babylonian "Epic of Creation" – *Enuma elish*, as it was anciently known from the opening words of the poem – dates from 1750 to 1400 BC, and describes the birth of the gods and the battle between Marduk, the chief god of Babylon, and Tiamat, with her army of monsters. After defeating Tiamat, Marduk returns to the falling body, and out of this creates the heavens:

> He rested, the lord, examining her body:
>
> Would divide up the monster, create a wonder of wonders!
> He slit her in two like a fish of the drying yards,
> The one half he positioned and secured as the sky...
> Therein traced he lines for the mighty gods,
> Stars, star-groups and constellations he appointed for them."
> He determined the year, marked out its divisions,
> For each of the twelve months appointed three rising stars...
>
> ***
>
> The eleven monster-(species) which Tiamat had created,
> Whose weapons he had broken, binding them at his feet,

[31]Neugebauer, *The Exact Sciences in Antiquity*, p. 48.

> He made of them (stellar) images to watch over the Apsu,
> That as signs (of the Zodiac) the group should ne'er be forgotten.[32]

Eliade astutely observes that in constructing a house or a temple or a city, man has always seen himself as imitating the paradigmatic creation of the gods, the cosmogony. Thus he finds the architectural plan of the temple "the work of the gods.... Through the grace of the gods, man attains to the dazzling vision of these models, which he then attempts to reproduce on earth."[33] Furthermore, we find: "The Babylonian king Gadea saw in a dream the goddess Nibada showing him a tablet on which were written the names of the beneficent stars, and a god revealed the plan of the temple to him. Sennacherib built Nineveh according to 'the plan established from most distant times in the configuration of the Heavens.'"[34] Though the idea of the heavens as an orderly system – or *kosmos* – is usually said to have begun with the Pythagoreans in the sixth century BC, clearly the kernel of the idea – if only a kernel – is already present in these far older Mesopotamian intuitions.

If the first constellations of the Zodiacal Quartet were, if Gurshtein is to be believed, already worked out as early as the middle of the sixth millennium, most of the other constellations apparently came into being between about 1300 and 1000 BC, or within a century or two of when the "Epic of Creation" was committed to baked clay tablets (to put this in perspective, this was broadly the period when Akhenaten, the "heretic king," reigned in Egypt, when Moses was leading the Israelites out of Egypt, and the Greeks were laying siege to the Anatolian city of Troy). As we have noted, it was the sexagesimal Babylonians who were responsible for dividing the ecliptic (the Sun's path across the sky) into 12 equal zones of celestial longitude, thereby forming the first known celestial coordinate system, and they also introduced the constellations of the Zodiac which at first seem to have only consisted of a few stars designated within each of the 12 equal zones. Indeed, in the passage quoted above, the Zodiacal constellations are related to the monster-species created by Tiamat – a *zoo-diac*, or circle of animals, indeed. The Mesopotamian constellations were later introduced into Greece and from thence they were assimilated, like so much else of the Mesopotamian culture, into the common heritage of humankind.

Among the Babylonian astronomer-priests – whose multiple roles included those of minding the gods, omens, and calendars – the stars came to be enveloped in (or still retained) something of the magic of earlier, Shamanistic, traditions.

---

[32]According to the note in D. Winton Thomas, ed., *Documents from Old Testament Times*, p. 14: Apsu was the great "male" ocean, especially the large tidal lake which anciently surrounded the city of Eridu in Southern Babylonia – consisting of marshes and waters of the subsoil. The word comes from the Sumerian Abzu, the first element being *ab* ("sea"); it appears in Greek as *abussos*.

[33]Eliade, *The Sacred and Profane*, p. 59.

[34]Ibid.

Magic rituals were often associated with the Sun, Moon, planets, and constellations. We learn, for instance, that Sargon the Great, of Akkad (reigned ca. 2350 BC), appointed one of his daughters priestess of the Moon in Ur; an evocative fact dating back to the time when the Old Testament patriarch Abraham was living there. (How vast are the depths of prehistory reaching back to long before the oldest memories of Genesis!) We learn that the stars, "the brightly shining gods who are judges," served as mediators between gods and men. However, they did not yet influence *everything* as they would afterwards come to do in astrology. Rather, men diffidently asked their advice, and sought from them portents or protections from evil.

Perhaps Sargon's daughter, priestess of the Moon in Ur, was responsible in her day for noting the first appearance of the crescent Moon, an event marking the beginning of the lunar month. Even before her time, Sumer-Akkadian star-watchers were struck as people have been in all times and places by the brilliant planet Venus, which often appeared in the same field of observation with the Moon. The almost-new Moon standing together with Venus in the sky is one of the most impressive sights of naked-eye astronomy, and seems to have been recorded in one of the earliest known monuments of the Sumerians, the Ur-Nammu stele from the reign of Ur-Nammu, the builder of the great ziggurat of Ur (ca. 2113–2096 BC).[35]

Venus and the Moon cast their romance as they converge in the morning sky. Based on the observations by Julian Baum at Beeston Castle, Chesire, England, on December 1, 2008. Painting by Julian Baum. © Julian Baum

[35]S. N. Kramer, *The Sumerians*. Chicago: University of Chicago Press, 1963; illustration after p. 64.

In this stele, the star is 12-pointed. However, on later Sumerian seals a crescent-Moon is shown in the embrace of an eight-pointed star. There is little doubt that the latter is meant as a symbol of Venus, where the eight points recall the planet's intricate 8-year period (see below).[36]

The Sumerian name for Venus was Inanna. She was originally a rain-deity and fertility-goddess, bride of the god Dumuzi-Amaushumgalana, who represented the growth and fecundity of the date-palm (hence she was sometimes known as the "Lady of the Date Clusters"). She set her heart on ruling the Underworld and attempted to depose her sister (Ereshkigal, Lady of the Greater Earth), but the attempt failed. She was killed and dispatched to the Underworld. Eventually, Enki, the Lord of Sweet Waters in the Earth, managed to bring her back, but only on condition that she offer a substitute in her place. She chose her husband Dumuzi, when she found him feasting instead of mourning her absence. In the end, Dumuzi and his sister Gesthtinanna were allowed to alternate as her substitute; each spent half a year in the Underworld and half a year above it.

Innana is the supreme goddess of the ancient Sumerian cuneiform temple-tablets. Later she would be identified as Ishtar or Astarte, whom the English poet John Milton would recall as

> Astoreth, whom the Phoenicians call'd
> Astarte, Queen of Heav'n, with crescent Horns;
> To whose bright Image nightly by the Moon
> Sidonian virgins paid their vows. and songs.[37]

In Egypt, she was Isis – not Venus but the goddess of the Dog Star, whose annual appearance in the sky announced the earth-fructifying flood season of the Nile.

Undoubtedly the ancient Sumerian story of Innana's descent into and return from the nether world – the "land of no return," the realm of death and darkness – is a principal source of the Babylonian, Assyrian, Phoenician, and Biblical traditions. Thus we read (in the translation of Samuel Noah Kramer):

> Innana ascends from the nether world,
> The Anunnaki fled,
> And whoever of the nether world may have descended peace-
> fully to the nether world;
> When Innana ascends from the nether world,
> Verily the dead ascend ahead of her.[38]

---

[36]Fevzi Kurtoglu, *Turk Bayragi ve Ay Yildiz*. Ankara: Truk Tarih Kurumu Yayinlari, 1992, p. 25.

[37]*Paradise Lost*, I, 437–440. She was also referred to as the "torch of heaven." Sir Henry Layard found a representation of Astarte at Nineveh, bearing a staff tipped with a crescent.

[38]S. N. Kramer, *Sumerian Mythology*. American Philosophical Society Memoirs, vol. XXI. Philadelphia, 1944, pp. 86–93.

Fertility rites and resurrection myths inevitably became entwined in the accretive mythological imagination around the planet Venus's alternate appearances above the horizon as the Morning Star (where it reigns for 260 days; close to the period of human gestation) followed by its long disappearance into the Underworld and rebirth as the Evening Star. In addition to its bobbing motions east and west of the Sun, which carry it alternately into the evening and morning sky, Venus pursues the Sun as it moves north and south with the seasons. The planet runs ahead of the Sun along the Zodiac by as much as a month and a half when it is in the evening sky, then falls behind it by as much as a month and a half when it is in the morning sky.

Venus's actual observed motions consist of a series of intricate dance-steps that trace out complex, but characteristic, regularly repeating curves and loops.

Because Venus takes 225 days to orbit the Sun, and the Earth, orbiting in the same direction, takes 365 days, after each elongation the Earth falls behind Venus. Venus, like a faster racehorse, catches up again, and the two planets return to the same positions relative to the Sun after 584 days. This period marks the interval between one appearance at inferior conjunction and the next. (Alternatively, between successive superior conjunctions, greatest elongations east, or greatest elongations west of the Sun.) This is known as the *synodic period.*

There happens to be something truly remarkable about this figure, as the numerology-obsessed Babylonians discovered: $584 \times 5 = 2{,}920$, but also $365 \times 8 = 2{,}920$! To put it another way: whatever Venus is doing tonight, it will be doing exactly the same thing 8 years from tonight.

This must have seemed as marvelous to whatever unknown Babylonian priest-astronomer made the discovery as the Pythagorean theorem would one day seem to Pythagoras of Samos. That unknown Babylonian had realized that, as we should now say, there is a 5/8 orbital resonance between the Earth and Venus. It was known already by (at latest) the middle of the second millennium BC. We know this from the remarkable "Venus tablet," now preserved in the British Museum. This tablet once formed part of the library of the Assyrian king Ashurbanipal (670–630 BC). It contains a record of the appearances and disappearances of Venus – referred to as Nin-dar-anna, "mistress of the heavens" – during the 21 year reign of King Ammisaduqa, the next-to-last king of the first Babylonian dynasty. It used to be thought that the observations in the Venus tablet pointed to a specific date for the years of Ammisaduqa's reign; if it had, it would have securely anchored the whole chronology of the ancient kings of Babylon. Unfortunately, the dates remain uncertain, though it does seem certain that Ammisaduqa reigned between 1650 and 1550 BC. Apart from the dates of the appearances and disappearances of Venus, the text is mainly concerned with the interpretation of various omens such as:

> In the month Abu on the sixth day Nin-dar-anna appears in the east; rains will be in the heavens, there will be devastations. Until the tenth day of Nisannu she stands in the east; at the eleventh day she disappears. Three months she stays away from the heavens; on the eleventh day of Duzu Nin-dar-anna flares up in the west. Hostility will be in the land; the crops will prosper.[39]

[39]A. Pannekoek, *A History of Astronomy*. New York: Dover, 1989 reprint of 1961 ed., p. 33.

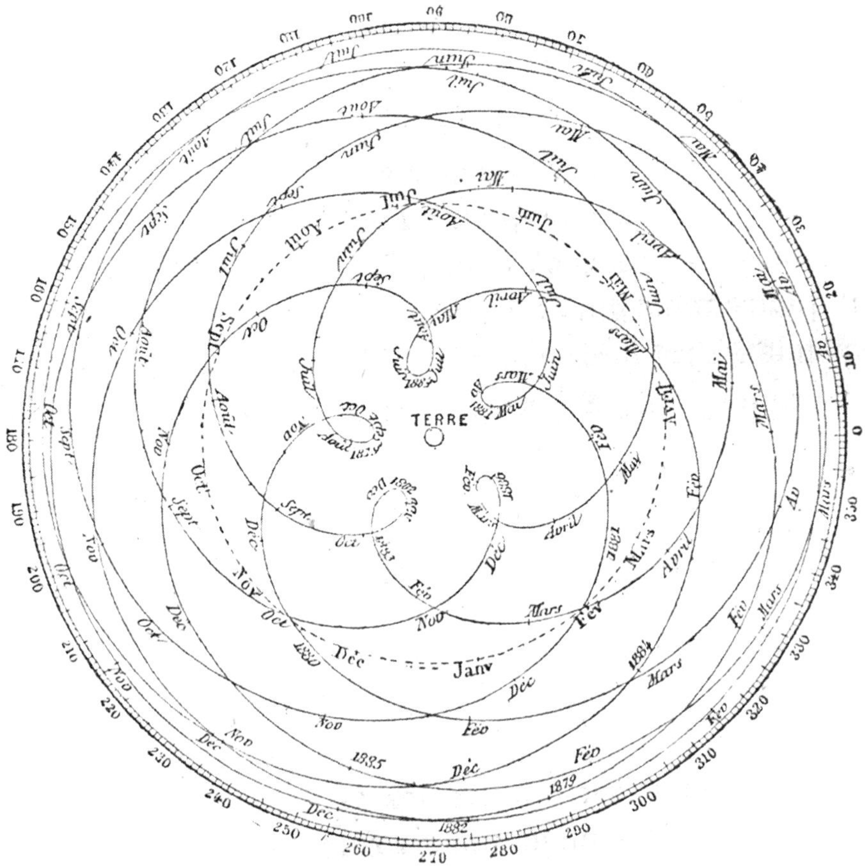

Apparent motion of Venus. The 5:8 synchrony between the orbital motions of the Earth and Venus, marveled at by the Mesopotamian astronomers in antiquity, is illustrated in the rosette-like figure traced by Venus relative to the Earth between 1879 and 1887. Diagram by Camille Flammarion, *Les Terres du Ciel*

These comments would seem to imply that precise positions of Venus's appearances on the horizon were thought to correlate with such life-and-death matters as cycles of rain and drought, of planting and harvest. Ammisaduqa and the other Babylonian kings clearly consulted the astronomer-priests to find out what Venus had to say about the best times for making peace or waging war. (Ironically, his own reign ended with a devastating raid from the neighboring Hittites; within only 20 years, control of the city would pass to a new dynasty.)

I began to record my own observations of Venus in the spring of 1964, some 36 or 37 centuries after the observations recorded on the Venus tablet during the reign of Ammisaduqa. For more than 40 years I have continued to passionately follow the rhythms of the planet's alternate appearances in the evening and morning sky and

The celebrated Venus tablet, catalog no. K. 150 in the British Museum. Found by Sir Henry Layard at Kouyunjik (ancient Nineveh) among the remains of the library of the eighth century Assyrian king Ashurbanipal, they appear to record Venus observations from the time of King Ammasiduqa in the seventeenth century BC. Photograph by William Sheehan in 2004

its periods of absence in between. Below I list a series of dates of all the elongations east (Venus an evening star) during the entire period covered by these observations, and have highlighted in boldface a series illustrating the 8-year cycle:

The 8-year Venus-solar period is not quite perfect; it falls short by a little more than 2½ days, which means there is a slight slippage of alignment. As is apparent from the

Year	Greatest elongation, E.	Inferior conjunction
1964	**April 10**	**June 19**
1965	Nov. 15	
1966		Jan. 26
1967	June 20	Aug. 29
1969	Jan. 26	Apr. 8
1970	Sept. 1	Nov. 10
1972	**Apr. 8**	**June 17**
1973	Nov. 13	
1974		Jan. 23
1975	June 18	Aug. 27
1977	Jan. 24	Apr. 6
1978	Aug. 29	Nov. 7
1980	**Apr. 5**	**June 15**
1981	Nov. 11	
1982		Jan. 21
1983	June 16	Aug. 21
1986	Aug. 27	Nov. 5
1988	**Apr. 3**	**June 13**
1989	Nov. 8	
1990		Jan. 19
1991	June 13	Aug. 22
1993	Jan. 19	Apr. 1
1994	Aug. 25	Nov. 2
1996	**Apr. 1**	**June 10**
1997	Nov. 1	
1998		Jan. 16
1999	June 11	Aug. 20
2001	Jan. 17	March 30
2002	Aug. 22	Oct. 31
2004	**Mar. 29**	**June 8**[a]
2005	Nov. 3	
2006		Jan. 13
2007	June 9	Aug. 18
2009	Jan. 14	March 27

[a]At inferior conjunction on June 8, 2004, the planet passed directly between the Earth and the Sun, and appeared as a dark spot against the disk of the Sun, a phenomenon known as a transit.

table, the planet comes to Greatest Elongation East of the Sun 2 or 3 days earlier in the season after 8 years. In other words, the 5/8ths orbital resonance is not quite exact.[40]

If the Earth were indeed the center of the motions of the planets, as used to be believed, then the motions of Venus would resemble a lovely rosette-like pattern (technically known as an *epitrochoid*).

[40] Because Venus's orbit is tilted slightly to that of the Earth, usually Venus passes either slightly above or slightly below the Sun at inferior conjunction. This is why transits of Venus are so rare. The 8-year period is the reason why transits of Venus occur 8 years apart. Thus the June 8, 2004 transit will be followed by another transit on June 6, 2012. One can also understand why such a long interval (more than a century) must pass between pairs, since a transit can occur only when Venus is near one of the two nodal points of its orbit and it takes that long before Venus and the Sun align again near a node.

In addition to the Venus cycle, there is also an 8-year solar-lunar cycle: eight solar years contain almost exactly 99 lunations. Thus we have:

8 solar years = 2,921.937592 days
99 lunations = 2,923.52841 days
5 Venus synodic cycles = 2,920 days

What this means is that the three most important heavenly bodies – the so-called "heavenly triad" of the Sun, Moon, and Venus – are involved with one another in an intricately choreographed celestial dance.

This 8-year period was known as the "Great Year," and is reflected in the myth about the Cretan King Minos, who was supposed to have reigned for 8 years. In the ninth year, he descended into the cave of his father Zeus to bring back a new set of laws. The myth, which is thought to reflect the interval between occasions on which the midwinter solstice occurs close to the time of new moon, contributes elements to the famous story of the Minotaur in the labyrinth. It is possible that the labyrinth was a true maze, marked out in mosaic on a pavement as a ritual dancing pattern as seems to be recalled by Homer (*Iliad* xviii, 592):

Daedalus in Cnossus once contrived
A dancing-floor for fair-haired Ariadne.

If so, it must have been meant, at some level, as a representation of the mazy movements of the planets.

The point is that early on – by the second millennium BC – humans were finding meaningful connections between their lives and the rhythms of the Sun, Moon, planets, and stars. They counted days, months, and years, and began to work out the periods of the planets. Among the Babylonians especially, arithmetical computations using a sextagesimal system, with the numbers written out in cuneiform on clay tablets, reached a surprisingly high degree of sophistication. Much of this number-lore – including the whole-number ratios in the periods of Sun, Moon, and planets – later came into the hands of the Greeks, as will be discussed in the next chapter.

As I write this (September 2008) I am eagerly anticipating Venus's next apparition as a brilliant interloper into the evening sky – one of the events that first sealed my fate as a passionate amateur astronomer many years ago. Like Innana, I have descended into the depths, to summon something from my personal past and also from the deep past of Western civilization.

Forty thousand years ago, the ancestors of the cave painters had arrived in Europe and were led west from Africa (which they entered via the Middle East) to France and Spain. It may well have been they were led by the Evening Star.

Four thousand years ago, the pirouettes of Innana were being wondered at by the peoples of the civilization of the Fertile Crescent in the land we now know as Iraq.

Four hundred years ago, Galileo first imaged the Evening Star with his telescope.

Forty years ago the planet – lovely Venus-Aphrodite-Ishtar – awakened my own imagination, and inspired my passion for the planets.

The cycles of Venus have scribbled their way across the lifetime of our kind – interleaving the turning pages of biography and history and sustaining passions for worlds beyond. And so they shall do as long as there is something in humans that resonates to the orderly movements of the planets.

Chapter 5
Pure Ambrosia

With Centric and Eccentric scribbl'd o'er,
Cycle and Epicycle, Orb in Orb.

Milton, Paradise Lost, VIII, 83–84.

As the Venus tablet shows, omens were the original focal-point of the ancient Mesopotamian astronomers. These omens are collected in the so-called "Enuma Anu Enlil" tablets,[1] a series of 70 tablets containing thousands of omens dating back to the second millennium BC. For instance: "If Jupiter [rises] in the path of

Marduk (Jupiter) and the Milky Way. Photograph by William Sheehan, 2008

[1]The name means "when Anu and Enlil …," from the opening words of the first tablet. Anu and Enlil were Sumerian gods.

W. Sheehan, *A Passion for the Planets: Envisioning Other Worlds, From the Pleistocene to the Age of the Telescope*, DOI 10.1007/978-1-4419-5971-3_5,

the [the god Enlil's] stars, the king of Akkad will become strong and [overthrow] his enemies in all lands in battle."[2]. Though many are celestial omens, others record meteorological phenomena – a not insignificant point, as we shall see.

Among the ancient Sumerians and Babylonians, the planets were not actual gods; rather they were *manifestations*, interpreters, of the gods, the "stars of the great gods who ruled the world." Their wandering among the background stars, reverses of direction, or combinations with the Moon, other planets and stars provided cryptic commentary on terrestrial affairs. Thus arose that majestic and fatalistic vision of reality that would have such a profound influence on the course of later civilizations, and that seemed to find its grandest expression in the alternating rise and wreck of empires:

> When I consider every thing that grows
> Holds in perfection but a little moment;
> That this huge stage presenteth nought but shows
> Whereon the stars in secret influence comment;
> When I perceive that men as plants increase,
> Cheer'd and check'd even by the self-same sky,
> Vaunt in their youthful sap, at height decrease,
> And wear their brave state out of memory.[3]

The planetary commentary of the heavens was of course in code; it had to be deciphered. The magic art – or presumption on human credulity – of being able to do this is the subject-matter of astrology. The Babylonians were astrally oriented but not in any modern sense; they were astronomers secondarily, and first and foremost "astrologers … and soothsayers."[4]

Parenthetically, I would add that astrology – and its cousin numerology – involved ideas that could only occur in a literate culture. Images make an immediate appeal to the beholder. True, they may benefit from captioning – from contextual annotation and explanation – but they do not wholly or even much depend on it. As with the communication we experience with animals, images speak to us through nonverbal means, through a diction of silence that can be as eloquent as any speech. No doubt the late-Pleistocene hunter-nomads who painted the caves of southern France and northern Spain told stories around their hearths, and they must have had some form of music. Indeed, it now seems well established that even the Neanderthals had language. But all that died away in the air of the Mammoth Steppe. Away from the hearth, the hunter was a stalker of animals, and though it was necessary for the hunter to sometimes communicate with his confederates this too must have occurred largely through non-verbal means – sign language and the

[2] N. M. Swerdlow, *Babylonian Theory of the Planets*. Princeton, NJ: Princeton University Press, 1998, p. 94.

[3] Shakespeare, Sonnet 15.

[4] Daniel 5:7.

like; a too garrulous hunter would be an assured failure! Hunting has a solitary and silent aspect; it relies on acute senses – hearing, touch, smell, but chiefly vision.

On the other hand, civilized society – social relations among large populations – depends mainly on the spoken and (increasingly) the written word. "In the beginning was the word" applies only to the post-Neolithic era. Before that was the image, which goes back much farther than the word.

What happened with the invention of writing (and it seems to have followed the same process in all the societies in which its invention occurred) was that figures – pictographic symbols – were first used as reminders of ideas or facts, but later came to be associated with a series of marks (such as those in cuneiform) which came to stand not for things themselves but for syllables – sounds – of speech. The Sumerian script, says historian of science George Sarton, was based on the use of some 350 syllabic sounds, and could be – and was – adapted to the expression of a host of languages "as unrelated as Sumerian on the one hand and the Eastern Semitic dialects on the other. Nor was it restricted to the peoples of Mesopotamia. It spread to the countries east of the Tigris and to those north and west of the two rivers."[5] The advantage of the written word was that it allowed "the graphical representation of abstractions, or feelings, or proper names."[6] The abstractions included numerical data and proper names those of kings, the cities they ruled, the tribes they fought, and the tribute they gained in conquest. The disadvantage – or advantage, depending on how one looks at it – is that the signs are completely arbitrary – they are no longer pictographic but instead form a code, and do not immediately reveal their meaning. The study of arbitrary symbols is known as semiotics. In order to make sense of them, one has to know the code and decipher them (this, by the way, also involves the transfer of the locus of functioning from the Right Hemisphere of the brain, which is predominantly pictorial, to the Left, which is predominantly verbal).

If the planets were writing in the sky, obviously what they were writing was not self-evident. It was cryptic, in code. The planets were the interpreters of the gods, but their interpretations also had to be interpreted. The method the Babylonian astrologers devised to break the code was astrology. Admittedly, astrology is apt to seem rather foolish today; however, it led in fruitful directions, and mathematics itself, which has been the handmaiden of science, has very deep roots in cryptology and numerology.

Astrology certainly connected the life of man to the heavens as it had never been before. It opened his eyes, awakened his interest. It was not intrinsic interest in the phenomena of the sky but the imputed role of the planets as the "interpreters" of the will of the gods that first led the scribes, as early as the eighth century BC, to begin keeping meticulous records of them. These records are now known as the

---

[5] George Sarton, *A History of Science*, volume 1: Ancient Science Through the Golden Age of Greece. New York: W.W. Norton, 1970, p. 64.

[6] Ibid., p. 63.

"Astronomical Diaries" and record celestial phenomena corresponding to those in the omens.[7] Chief among them were records of eclipses and of the heliacal risings and settings of the planets (the times when a planet can first be sighted ahead of the Sun at sunrise or after the Sun at sunset). These records, maintained over centuries, would eventually lead to the discovery of regularities in the motions of the planets in the form of exact periods in which the planets returned to essentially the same positions again. Thus the planetary motions – however irregular they may appear at first – were found to be periodic. Henceforth even these "willful ones" were tamed, domesticated, reduced to a predictable order – rather, in fact, like the floodplains of the Tigris and Euphrates.

From data like that presented at the end of the last chapter, it was known already by the second millennium BC that Venus's movements follow an 8-year cycle, after which Venus returns nearly to exactly the same place, under the same circumstances (e.g., Greatest Elongation East, Inferior Conjunction, Greatest Elongation West, Superior Conjunction) in the sky. Jupiter – the Star of Marduk, the chief god of the Babylonian empire since before the time of Ammasiduqa and Hammurabi – was noted to make its heliacal risings, the point where it first emerges from the Sun before dawn, at intervals of roughly 399 days (the mean synodic period). However, each of these events is displaced one Zodiacal sign from the last, and the Jupiter-cycle – that in which the planet returns to the same position (heliacal rise or opposition) at the same point in the sky – is 71 years. Since the Earth, Sun, and planet are now again in the same relative positions as at the beginning of the cycle, the speeds of the Sun and Jupiter through the Zodiac will be the same as before. Thus the whole cycle repeats.

Eventually the scribes worked out similar periods of recurrence for all the planets, data that became pure gold when mined by Greek astronomers such as Hipparcos and Ptolemy for use in their planetary theories. This data (as given by Ptolemy in the second century AD but essentially Babylonian data) is summarized below (where s.p. = synodic period, y = years, d = days, and rev. = revolutions):

Saturn 57 s.p. = 59 y. + $1\frac{3}{4}$ d. = 2 rev. + $1^\circ\ 43'$.
Jupiter 65 s.p. = 71 y. + $\frac{49}{10}$ d. = 6 rev. − $4^\circ\ 50'$.
Mars 37 s.p. = 79 y. + $3\frac{13}{60}$ d. = 42 rev. + $3^\circ\ 10'$.
Venus 5 s.p. = 8 y. − $2\frac{2}{10}$ d. = 8 rev. − $2^\circ\ 15'$.
Mercury 145 s.p. = 46 y. + $1\frac{1}{30}$ d. = 46 rev. + 1°.

[7] According to Neugebauer, *The Exact Sciences in Antiquity*, p. 101: "Around 700 BC, under the Assyrian empire, we meet with systematic observational reports of astronomers to the court. In these reports no clear distinction is yet made between astronomical and meteorological phenomena. Clouds and halos are on equal footing with eclipses. Nevertheless, it had been already recognized that solar eclipses are only possible at the end of a month (new moon), lunar eclipses at the middle… We should recall here Ptolemy's statement that eclipse records were available to him from the time of Nabonassar (747 BC) onwards."

Once they were in possession of data of this sort, the scribes – by the third century BC – were able to produce ephemerides: tables (or rather tablets) containing computed positions of the Sun, Moon, or planets for intervals of a day, or a month. Historian of science Noel Swerdlow, who has worked out in detail the way the Babylonians arrived at the parameters adopted in computing these ephemerides, suggests that the reason the scribes turned to the computation of these phenomena rather than relying on direct observations of them was owing to the frequent interposition of bad weather that prevented them from observing crucial celestial events:

> All those nights of rain and clouds and poor visibility reported in the Diaries turned out to be good for something after all. When it is clear, observe; when it is cloudy, compute.... Here ... science was born in Babylon. From bad weather was born good science. And the reduction of periodic natural phenomena, however great their irregularities, to a precise mathematical description that may be applied to both prospective and retrospective calculation, that is, to mathematical science, was ... the achievement of the Babylonians.[8]

Incidentally, the supposed transparency of the Mesopotamian skies has often been invoked to account for marvelous observations, such as the naked-eye perception of the crescent of Venus sometimes claimed for the ancient Babylonians. But it is a myth. Recent experience has shown that the sky in modern Iraq (as in other deserts) is usually veiled in dust. It must have been so even in antiquity.

As the priest-astronomers of Sumer-Akkadia, Babylon and Assyria were growing steadily more conversant with the motions of the planets, their Greek neighbors in Asia Minor and the Peloponnese long lagged behind. The Neolithic Sumerians who invented writing on baked tablets at first specialized in drawing up lists; so the Bronze Age Myceneans – they who, the poets later claimed, launched a thousand ships to bring back the most beautiful woman in the world from Anatolia – recorded in the Linear B script on their tablets mostly mundane details. The Linear B script (deciphered only as recently as the early 1950s by Michael Ventris) was utilized primarily for bureaucratic purposes such as drawing up tallies of wine, pottery vessels, grain, oils and livestock, the assets of the material culture that warrior-overlords controlled. Our disappointment at the mundaneness of these records, as with the majority of those kept by the Babylonians, is keen. Archaeologist Bettany Hughes points out: "In the tablets dug up to date there is little that is immediately recognizable as the inner voice of a civilisation, no self-conscious historical record. This is not a culture that employed written symbols as a form of emotional expression."[9]

The center of this culture was Mycenae, on the Peloponnese, and its ways Homer recalls sometimes vividly, more often dimly; it was a civilization dark, ruthless, and bloody, bristling and armed to the teeth. While the Babylonian astronomers

---

[8] Swerdlow, *Babylonian Theory of the Planets*, p. 56.

[9] Bettany Hughes, *Helen of Troy: goddess, princess, whore*. New York: Alfred A. Knopf, 2005, p. 8.

were recording omens, the Myceneans, who had the souls of swashbuckling buccaneers, were building fortresses along trade routes – like Mycenae itself – with commanding Cyclopean walls, reinforced ramparts, and crouching positions beneath menacing mountains. But by the twelfth century BC – shortly after the Greeks sacked Troy – their world was in pieces. They lost the secret of writing for centuries; indeed, Linear B itself would never be used again. It would remain mute – because undecipherable – for 3,000 years. Across the Dardanelles, the once-mighty Hittite empire, which had connections with the Myceneans, the Babylonians, and the Egyptians, was also crumbling; many of the Bronze-Age sites appear to have been struck repeatedly during this period by powerful earthquakes. One has the sense of a world undergoing "systems collapse."

Nevertheless, a tribal memory survived. From the distance of centuries the oral poets – Homer and Hesiod and their ilk, who were the tribal encylopaedias of the Greeks until the reintroduction of written language – were free to embellish what they remembered, for they were not tied as literal societies are to written records with their unforgiving literalness. Perhaps after all it was the loss of literacy in the interim that gives the *Iliad* and the *Odyssey* the qualities that the cuneiform tablets – and the Babylonian Creation myth and Gilgamesh in even the most sympathetic reading – are found to lack.

The heroic Greeks as Homer recalls or recreates them are warriors, traders, sailors, and pirates. They still use the sharpened spear, but in contests of skill and wits with other men rather than with Beasts as in the Ice Age, as they fight over spoils and women. They hope to cover themselves by such means in glory. One sees the image and the ethos of the hunter – but changed and intensified in the close quarters of civilization; the plain before Troy is a postage-stamp version of the Mammoth Steppe. A vast fatalism hangs over the warring heroes. One knows their dooms have already been pronounced by the gods. The end is death. Humans are "miserable mortals who, like leaves, at one moment flame with life, eating the produce of the land, and at another moment weakly perish."[10]

It is a tragic vision indeed but one that is not to be read in the stars. Unlike the Babylonians, the Greeks were not astrally-oriented. One would have hardly guessed that this warlike people who (if Homer is to be trusted) knew the names of only a few stars would take the next great step forward in astronomy.

In the *Iliad*, Homer mentions the names of a few specific stars and star-groups – Orion, Sirius ("the star which rises in late summer"), the Pleiades, the Hyades. In the *Odyssey*, he mentions a few others. But that is all.

He refers to the Evening Star (Hesper) and the Morning Star (Phosphor) as if they were separate objects – something that had already long since been shown to be incorrect by the Egyptians and Babylonians. The first Greek to proclaim the identity of the Evening and Morning Stars appears to have been Pythagoras in the sixth century BC,[11] the same century in which the Homeric epics were committed

[10]*Iliad*, XXI, l. 463.

[11]Others, however, claim that Parmenides was the first to do so.

to writing. Pythagoras, as we shall see, made many original discoveries. However, as he is supposed to have spent 22 years studying in Egypt and 12 years in Babylon, he must have learned of the identity of the Evening and Morning Stars there.

The Homeric epics were responsible for emphasizing the importance of the individual – we have here the first fully individualized characters in literature – and what seems to be related to it, an ethos of intense competition. The Greeks and Trojans are individuals engaging each other *mano-a-mano* for glory. In an intriguing book, *Soldiers and Ghosts*, about the way that Homer's descriptions of heroic one-on-one fighting affected the later Greek ideas of warfare, J. E. Lendon writes:

> The heroic one-on-one fighting described in the *Iliad* is, as has long been understood, closely linked to the heroic motives the poem attributes to the fighters. Homeric heroes compete with each other and are conceived as being ranked one against another in a competitive series, each yearning "always to be the best and preeminent above others." … Nearly every activity in the *Iliad* can be imagined to be a competition…. In the epic formula, battle is "where men win glory."[12]

It may be true, as Swerdlow says, "the discussions of two Scribes of Enuma Enlil contained more rigorous science than the speculations of twenty philosophers speaking Greek…. The origin of rigorous, technical science was not Greek but Babylonian, not Indo-European but Semitic…. and, my God, those Scribes were smart."[13] Smart as they were, however, there was no glory in being smart; the Babylonian scribes are nameless to history. Theirs is a collective effort; they are unnamed servants – diligent, hardworking bureaucrats – and they belong to the king. The kings alone have names. But then the kings are not men but gods on Earth.

By contrast, Greek science was created by a series of individuals. From the first awakenings of Ionian speculation among the so-called pre-Socratics, such as Thales, Anaximander, and Pythagoras, the names of individual men will be attached, like personal property, to their individual ideas. (They competed with one another by making sweeping statements such as "All is water," "all is air," "everything changes," "nothing changes," "all is number." Nothing could be further from the Babylonian ways of thought in which abstractions – and general laws – were avoided.)[14] The pre-Socratics were clearly men who were in keen intellectual competition with one another. They battled one another in mind as the military heroes of the *Iliad* did in body; like the Homeric heroes, battle was for them "where men win glory."

---

[12] J. E. Lendon, *Soldiers and Ghosts: a history of battle in classical antiquity*. New Haven and London: Yale University Press, 2005, p. 24.

[13] Swerdlow, *Babylonian Theory of the Planets*, pp. 181–182.

[14] Even less was there any similarity to the great men of the Hebrew culture developing in Palestine which, though closer to Ionia than to Egypt or Mesopotamia was – as George Sarton points out – "as foreign to the Greeks as either of those, if not more so." Sarton adds in *A History of Science: ancient Greece through the Golden Age of Greece* (New York: W.W. Norton, 1970), vol. 1, p. 163: "By the end of the seventh century many of the prophetic books of our Bible had already been composed: Amos, Hosea, Micah, Isaiah, Hezekiah, Zephaniah, Jeremiah, Nahum, Habakuk; the

Pythagoras was born in the sixth century BC on the Ionian island of Samos. We have only the rudiments of a biography; he appears as a wonder-working mathematician whose being is shrouded in mystery and legend.

Pythagoras' birthplace, Samos, is one of the small volcanic isles that lie in the Aegean off the Anatolian coast (now western Turkey), where the Greeks established vigorous and thriving colonies and pursued trade which brought them constantly into close physical and intellectual contact with other cultures. As members of the mercantile class grew wealthier as a result of this activity, they could begin to afford to educate their sons. Richard E. Nisbett, in *The Geography of Thought*, argues that

> The drive toward education was apparently the result of curiosity and a belief in the value of knowledge for its own sake. The curiosity characteristic of the Greeks may in turn be explained in part by the location of the Greeks at a crossroads of the world. They were constantly encountering novel and perplexing people, customs, and beliefs.[15]

Instead of retreating from ideas and customs that differed from their own, they found the clash of ideas exciting and quickly cultivated a penchant for debate which was quite unprecedented and radically different from anything that had appeared in the authoritarian and tradition-bound societies of Egypt or Mesopotamia.

Pythagoras is said to have left Samos to escape the tyranny of its ruler, Polycrates. According to one tradition, Pythagoras went first to Miletus where he became associated with Thales and Anaximander, then to Egypt where, as noted above, he is supposed to have spent 22 years studying astronomy, geometry and the mysteries of the Egyptian religion (and must have learned of the identity of the Evening and Morning Stars). Then followed 12 more years before returning to Samos again. On finding only a single disciple, he resumed his wanderings and finally arrived at Croton, in southern Italy. The Crotonians found him rather more to their taste; his lectures were thronged, and he established an influential school.

Pythagoras must have possessed a remarkably syncretistic mind, and the ideas associated with him have always seemed a remarkable mélange of rational ideas and mystical intuitions. Ionia in the sixth century BC was a materialistic culture; but a spiritual revival was sweeping the mainland involving the worship of the god Orpheus (who originally came from Thrace). The Orphics were intellectualized successors to the Maenads, who worshipped the god Dionysos. They sought,

---

Pentateuch (or Torah), and the books of Samuel were already completed.... Let us ... consider now only the Prophets and the Torah, and compare them with the Homeric writings. The difference between the respective languages, Hebrew and Greek, is small as compared with that between ways of thought. The Hebrew prophet was a seer; the rhapsodist, a poet and storyteller. The latter referred sometimes to the gods and the heroes just as he referred to ordinary mortals, but the former spoke in God's name, in the name of the one God and of eternal justice. The contrast is so great that communication between the Hebrews and the Ionians was probably reduced to a minimum." One can understand for this why the roots of modern science are to be found among the Ionian Greeks and not among the Hebrews who remained scientifically illiterate. The amount of astronomical knowledge in the Bible, even for the time, is minuscule.

[15] Richard E. Nisbett, *The Geography of Thought*. New York: Simon and Schuster, 2003, p. 31.

through wild orgiastic dances to the music of cymbals and flutes on mountain sides, to achieve "out of body" experiences, *ecstasies*, in which the soul revealed its true – divine – nature and achieved union with the god.[16] The Orphics taught that the soul retained a consciousness of its divine origin – an "intimation of immortality," as it were – but was imprisoned or entombed in the body and could only be helped back to its divine state through the strict observance of taboos, penalties, and purifications. Many ideas from the Orphics would be accepted or adapted by Pythagoras and his followers.[17] But what was uniquely Pythagorean was his discovery of a new route to ecstasy. He reached it through contemplation of mathematical truths.

Pythagoras stands at a crossroads between two worlds. It is not much of an exaggeration to call him both the last Shaman and the first scientist. Shamans from time immemorial have cultivated the art of inducing trances by means of rhythms; thus the Shaman is often depicted with his drum. Pythagoras had his musical instrument – not a drum in his case but a stringed instrument, probably a cither or a lyre. Such instruments are already mentioned in Homer, and had been in use long before in Babylonia and Egypt. Sometime in the century before Pythagoras, the legendary father of Greek music, Terpsander of Lesbos, is supposed to have increased the number of strings to seven (the mystic number associated by the Babylonians with the number of the planets). With such an instrument Pythagoras discovered that when a string is plucked it produces a set of harmonics. If you touch it at the nodal point – divide the string in half – you produce A, the "fundamental harmonic." If you subdivide the string again, so that the relationship is now 2:3, you produce the fifth; with 3:4, the fourth; with 4:5 the major third; with 5:6 the minor third.

Whether Pythagoras was a musician who dabbled in mathematics or a mathematician who was also a musician is unclear. We have only this from Aristotle: "Pythagoras son of Mnesarchos first worked at mathematics and arithmetic." The order doesn't really matter, however; there are deep-seated affinities between music and mathematics which are only now coming to be understood as reflections of a specialized neural organization in the brain.[18]

[16]See: E. R. Dodds, *The Greeks and the Irrational*. Berkeley: University of California Press, 1951.

[17]In these ideas of the sixth century BC, one recognizes vestiges of very much older ideas about man's state, including the "ecstatic journey" or vision-quest by means of which one might receive an understanding of the mystery of destiny and of existence after death. Those ideas had been current among the Shamans; for all we know they may have existed, in more or less developed form, in the Upper Paleolithic at the time of the cave painter. Eliade, in *Shamanism: archaic techniques of ecstasy*, p. 394, writes of the consistency of themes from the Shamans to the post-Pythagorean Plato: "The enormous gap that separates a shaman's ecstasy from Plato's contemplation, all the difference deepened by history and culture, changes nothing in this gaining consciousness of ultimate reality; it is through ecstasy that man fully realizes his situation in the world and his final destiny."

[18]For a fascinating discussion, see: Edward Rothstein, *Emblems of Mind: the inner life of music and mathematics*. Chicago: University of Chicago Press, 1995.

Pythagoras' first mathematical ideas seem to have been based on grouping pebbles in triangle and square associations. In this way he discovered various facts about odd and even numbers (triangular and square numbers according to Pythagoras' terminology). Possibly in part as a result of this method of calculating, numbers seemed more physical for him than they do for us, and he seems to have attributed physical properties to particular numbers and figures. He taught, for instance, that the cause of color was in the number five, the explanation of fire was to be discovered in the pyramid, etc. Clearly this was an idiosyncratic way of thinking. I can't help wondering if he mightn't have been one of those unusual people known as synesthetes – people who are able to associate numbers with shapes and colors in what seem to most of the rest of us to be very strange ways.[19] I have no way of proving it, but I would also find it hard to believe that he wasn't an autistic savant of some sort. What other kind of personality would possibly say, with any conviction, "All things are number"? Pythagoras seems to have recognized his own idiosyncrasy, for he is reported to have said: "There are men and gods, and beings like Pythagoras." He was also possessed in remarkable degree of *enthusiasmos* – another word with Orphic associations – and is said to have been so overjoyed on discovering his celebrated theorem concerning the right-angled triangle (the square of the hypotenuse is equal to the sum of the squares of the two remaining sides) that he immediately went out and sacrificed a hecatomb in thanksgiving to the gods.

Pythagoras eventually became mixed up with the local politics of southern Italy, came in on the wrong side, was exiled (again) and finally died, at Metapontium, about 497 BC. He had lived through to another troubled time; much of Ionia, including his native Samos, had been conquered by the expansionist Persians who – within a few years of Pythagoras' death – threatened Mainland Greece as well. They were turned back at Marathon in September 490 BC[20] by a combined force of Athenians and Plataeans – and notably without

[19]Such a person is the British savant Daniel Tammet. According to an interview by Richard Johnson in *The Guardian*, February 12, 2005: "Daniel Tammet is talking. As he talks, he studies my shirt and counts the stitches. Ever since the age of three, when he suffered an epileptic fit, Tammet has been obsessed with counting. Now he is 26, and a mathematical genius who can figure out cube roots quicker than a calculator and recall pi to 22514 decimal places. He also happens to be autistic, which is why he can't drive a car, wire a plug, or tell right from left…"

Tammet is calculating 377 multiplied by 795. Actually, he isn't "calculating": there is nothing conscious about what he is doing. He arrives at his answer instantly. Since his epileptic fit, he has been able to see numbers as shapes, colors and textures. The number two, for instance, is a motion, and five is a clap of thunder. "When I multiply numbers together, I see two shapes. The image starts to change and evolve, and a third shape emerges. That's the answer."

[20]The date of the battle was September 11, 490 BC which was the 17th day of the lunar month in Attica, according to N.G.L. Hammond in *The Cambridge Ancient History*, vol. IV (Persia, Greece and the Western Mediterranean c. 525 to 479), ed. John Boardman, N. G. L. Hammond, D. M. Lewis, and M. Ostwald, Cambridge: Cambridge University Press, 1988, p. 40. Hammond's calculations were made based on data provided to him by the Astronomer Royal, Sir Richard Woolley. Hammond notes that "the moon-goddess held an important place in the religious associations of the battle. The commemorative coins struck from 490 BC onwards showed a waning moon behind the owl of Athena…. Seltman and others have thought the waning moon was added to give the

Athenian tetradrachm showing the last-quarter Moon which figured in the events surrounding the battle of Marathon in 490 BC. William Sheehan collection

the assistance of the Spartans, whose arrival was delayed by their superstition that they must not march until after the Full Moon. The latter circumstance was famously memorialized on the Athenian silver tetradrachm which shows Athena on one side and her trademark owl – olive branch – and the waning Moon of Marathon on the other.

Pythagoras' disciples carried on after his death, espousing what must have been a rather "geeky" kind of religion. The Orphics had used the word "theory" to mean passionate sympathetic contemplation by a spectator of the suffering, death, and rebirth of their god, as reenacted in religious rites. The Pythagoreans seized on the word to mean the passionate contemplation of mathematical or cosmic truths. Here, for the first time, we meet men with a passion not, as among the Homeric heroes, for a beautiful woman or for the trophies and spoils of war or for the glory to be won in "man-slaughtering." Instead, they are passionate about numbers. This, at first blush, seems rather incomprehensible. Mathematics, music, and chess – three things by the way in which precocious achievements can sometimes occur even before adolescence, as George Steiner points out – seem to involve a certain specialized part of the brain

---

date of the battle.... It is more likely ... that the waning phase, shown on the coins, was a particular factor in the victory," perhaps by revealing that Persian cavalry, which had left its position, had not yet returned and allowing Militiades, the commander of the Greeks, "to thin his center, pack the wings and make its length equal to that of the Persian infantry line which was known could be attacked at speed across no-man's land before the cavalry could intervene."

which develops, it may well be, in isolation from the rest; a part of the brain that "sees the logical, the necessary harmonic and melodic argument as it arises out of an initial key relation or the preliminary fragments of a theme," so that "some chance instigation – a tune or harmonic progression … a row of figures set out for addition … the sight of the opening moves in a café chess game," can trigger "a chain reaction in one limited zone of the human psyche." In susceptible individuals, the result may be "a beauteous monomania."[21] This is something that is clearly kindred to our own passion for the planets, and is very different from the more typical passions for sex or dominance.

The fact that there are seven tones in the heptacord and seven planets impressed the Pythagoreans as something deep and necessary. They apparently asserted that the Sun, Moon, and five planets moved in circular paths in which their distances followed a musical progression. Thus the idea was born of a *kosmos*, an orderly system in the heavens governed by mathematical relations and harmonies. Pythagoras himself (according to an admittedly late tradition) is said to have affirmed that "the universe sings and is constructed in accordance with harmony."

In stating that "all is number," Pythagoras meant whole numbers or integers. But his ideas were already threatened soon after his death when one of his disciples discovered "irrational numbers"; those which cannot be represented as the ratio of one whole number to another.[22] Their discovery follows – ironically – directly from the Pythagorean theorem, and scandalized the Pythagoreans rather as the discovery of Neanderthal man did the pious Victorians. One of their brethren, Hippasos of Metapontium (c. 470 BC), later betrayed the oath of secrecy to which he was bound by revealing the secret of the dodecahedron, the last of the so-called regular solids to be discovered; he is supposed to have been punished – justly, in the view of his betrayed colleagues – by being drowned in a shipwreck. Perhaps it was Hippasos too who revealed the secret of the irrational numbers. In any event, the discovery of irrational numbers produced a crisis in Greek thought whose effects were to be felt for centuries; henceforth they discarded number and measurement – the only kind of mathematics the Babylonians had understood – and sought securer foundations in geometry.

Geometry proceeds by means of a series of philosophical conjectures – deductions from self-evident propositions (which are actually visuo-spatial intuitions). The "geometrical method" – the *esprit géométrique* as the French charmingly put

[21] George Steiner, "A Death of Kings"; in: *George Steiner: A Reader*, pp. 172–173.

[22] That discovery follows immediately from the Pythagorean theorem about triangles, a discovery which so elated Pythagoras that he is said to have sacrificed a pair of oxen in thanksgiving to the gods. The Pythagorean theorem states that in a right triangle, the square of the hypotenuse is equal to the sum of the squares of the lengths of the two sides. But the right-angled triangle in which the sides are each 1 unit in length, the hypotenuse is the square root of 2, a surd. In other words, it cannot be expressed as the ratio of two whole numbers, or is *ir*-rational. A simple proof, which is essentially that given in Euclid, Book X, is as follows. Let us suppose each side 1 unit long; then how long is the hypotenuse? Let us suppose its length is m/n units. Then $m^2/n^2 = 2$. If m and n have a common factor, divide it out, then either m or n must be odd. Now $m^2 = 2n^2$, therefore m^2 is even, therefore n is odd. Suppose $m = 2p$. Then $4\ p^2 = 2n^2$, therefore $n^2 = 2p^2$, and therefore n is even, *contra hyp*. Therefore no fraction m/n will measure the hypotenuse.

it – is well known from Euclid, and used to be taught in high schools as the model of rigorous rational, i.e., deductive thought. Though geometrical demonstrations originally must have involved manipulating objects such as lines, triangles and circles in diagrams sketched in the sand, and thus was empirical to a degree, eventually it came to represent a series of abstractions – *ideas* – that it seemed the mind was able to grasp directly, without dependence on the imperfect and unreliable senses. Thus in the teachings of Plato (427–347 BC), who was one of the most influential philosophers of all time, geometrical knowledge becomes intellectualized and disembodied as ideas existing forever in a pure Platonic heaven utterly divorced from experience. The senses are capable of deceiving and deluding; only ideal or abstract knowledge is divine and represents the ultimate reality. Plato introduces the persistent belief in what has been called the "mathematical myth," "the belief … in the explanatory and almost transcendent virtue of mathematics."[23] His views are most famously stated in his "Analogy of the Cave"; what is perceived by the senses are only shadows. Crucially, Plato believed that "the world [itself] is a copy, an

Plato. The bust in the Vatican Museum. Photograph by William Sheehan, 2008

[23]Jean-Pierre Changeux and Alain Connes. *Conversations on Mind, Matter, and Mathematics*. Princeton, NJ: Princeton University Press, 1995, p. 4.

image, of the real."[24] Thus any representation – be it in a work of art, or a model – is the copy of a copy, the image of an image. It could even be the copy of a copy of a copy, as when we admire a Roman copy of a Greek sculpture – such as the head of Plato here.

Plato's insistence that the intellect alone could grasp the truth was to be extraordinarily influential; eventually, any other view came to be seen as heretical. Plato turned against democracy after the defeat of Athens by Sparta in the hugely destructive Peloponnesian wars, which lasted from 431 to 404 BC. They were themselves an inevitable consequence of the "ferocious competitiveness of the Greeks" and the ideal of the warrior-hero Homer set up in the *Iliad*, which led to almost incessant and ultimately suicidal warfare among the city-states. As an aristocrat, Plato deeply distrusted commercial classes and tradesmen – one of whose representatives, Cleon, a tanner, had become the leader of the Athenian democracy after the death of Pericles during the Plague of 429 BC. Plato was also bitter after the execution of his teacher Socrates (in 399 BC) on charges of impiety and corrupting youth. In any case, his staunchly elitist attitudes were to ring down the ensuing centuries so that even as late as the second century AD, the Greek writer Plutarch, who "is continually running on the Aristotelian Ethics and high Platonic theories,"[25] could express the view that technical achievements of workmen and artists could be admired without admiring the men who produced them. Thus, says he:

> Many times, when we are pleased with the work, we slight and set little by the workman or artist himself... It was not said amiss by Antisthenes, when people told him that one Ismenias was an excellent piper. "It may be so," said he, "but he is a wretched human being, otherwise he would not have been an excellent piper."... Nor did any generous and ingenuous young man, at the sight of the statue of Jupiter at Pisa, ever desire to be a Phidias....[26]

Admittedly I am oversimplifying here, but many of the Greeks of the fourth century BC and later did take a rather anti-empirical approach; they distrusted all knowledge that rested on sensory experience or measurement, and insisted on starting only with abstract propositions and theoretical knowledge. In contrast to Babylonian astronomy, which was a matter of careful observation and arithmetical calculation, Greek astronomy became an extension of Greek geometry, in its methods and in its intuitions about space. Propositions – axioms – like those later codified by Euclid (c. 300 BC) were agreed. In addition the following two propositions were also regarded as axiomatic:

1. The Earth stands at rest at the center of the universe.
2. The heavens being a realm of perfection, the only motions allowed are simple uniform circular motions.

[24] F.M. Cornford, *Plato's Cosmology*. New York: Harcourt, Brace & Co., 1937, p. 28.

[25] Arthur Hugh Clough, Introduction to Plutarch, *Lives of the Noble Grecians and Romans* (trans. John Dryden and revised by Arthur Hugh Clough), New York: Modern Library, n.d., p. xxvii.

[26] Plutarch, "Pericles," in: *The Lives of the Noble Grecians and Romans*, translated by John Dryden and revised by Arthur Hugh Clough. New York: Modern Library, n.d., p. 183.

From Plato's time onwards, these became the rules of the game – a game rather like chess, and played with the chest master's passionate intensity. It provided "a focus for the emotions, as substantial as, often more substantial than, reality itself."[27] The sky was a chessboard, the moving pieces the planets.

It is said that Plato himself set before Eudoxos of Cnidus (c. 400 to c. 350 BC), the most able mathematician of the fourth century BC and a onetime student of Plato's Academy, the problem of "which uniform and ordered movements must be assumed to account for the apparent movements of the planets." If matching wits with the planets was a kind of chess, then Eudoxos was truly the first Grand Master.

By Eudoxos' time, the Greek astronomers recognized – as the Egyptians and Babylonians already recognized for a millennium or more – that there were basically two types of planetary motions needing to be explained: Mercury and Venus showed one type of motion, Mars, Jupiter, and Saturn the other. It is important to note that for naked-eye observers on the Earth, these apparent paths of the planets – projected onto the apparently flat surface of the sky – are the only data that are directly accessible. Albert Einstein has written:

> Since we human beings are tied to the earth, our observations will never directly reveal to us the "true" planetary motions, but only the intersections of the lines of sight (earth-planet) with the "fixed-star sphere."[28]

These intersections first began to be noted in the Neolithic. They were qualitatively well described by Eudoxos' time.

Thus Mercury and Venus are seen to rise above the horizon and fall back down (over the course of weeks or months, respectively) rather like projectiles cast by some giant atlatl. They never venture far from the Sun (Mercury never more than 28° from it, Venus never more than 48°), so that they are never to be seen in the deep dark night. They perform their dance exclusively in the evening or morning periods – which is why, ever since very early times, they had been known as evening or morning stars, and why the realization that these successive apparitions involved one and the same object was a first bold step that provided some kind of comprehensible structure to the universe.

Mars, Jupiter, and Saturn could be seen sometimes in the midnight skies, and their progress across the heavens was occasionally interrupted by backtracks or zig-zag movements – retrograde movements, so-called; rather like the move a Knight makes in chess. These reversals of motion did not occur at any time without rhyme or reason; they did so only around the times when these planets lined up opposite to the Sun in the sky. Mars moves more quickly than Jupiter and Saturn – indeed, at times it shifts its position by half the Moon's apparent diameter every night, and then its motion against the background stars is strikingly obvious. It also

[27]Steiner, "A Death of Kings," p. 173.

[28]Albert Einstein, Foreword to Galileo Galilei, *Dialogue Concerning the Two Chief World Systems*, translated, with revised notes, by Stillman Drake. Berkeley: University of California Press, 2nd revised edition, 1967, p. xv.

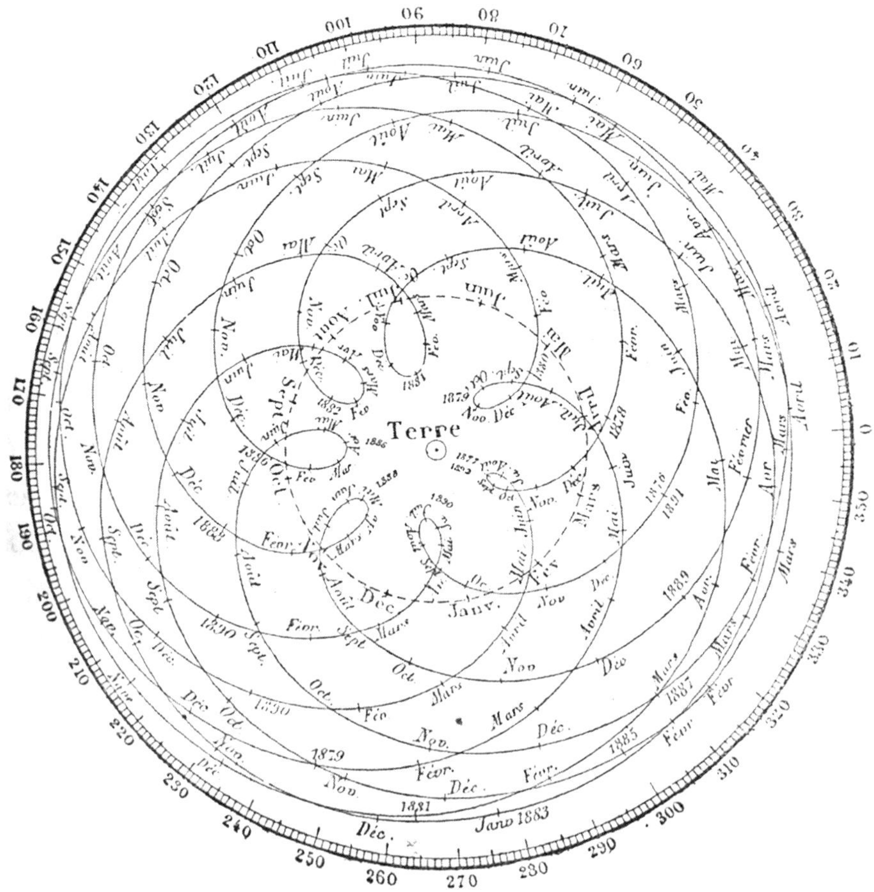

The apparent movement of Mars relative to the Earth, between 1877 and 1892. Diagram by Camille Flammarion, *Les Terres du Ciel*

undergoes dramatic changes of brightness of 50-fold. This made it the most quixotic of the planets, and early on set it apart as the astronomer's *bête-noire*. They engaged it – given its intense reddish hue, like that of a drop of blood, it was early identified with the god of war – in a centuries-long series of skirmishes and feints, attacks and counterattacks, in a kind of low-grade guerilla war.

Eudoxos was the first to devise a model to explain its retrograde motions. He surmised that Mars might be moving in a sphere upon a sphere rather like a wheel pinned to the rim of a larger wheel. Furthermore, the axis of Mars's sphere was tilted so as not to remain parallel to the axis of the other sphere, though the two axes remained connected to each other like a compass in a gimbal. Eudoxos's proposal has been described as the system of homocentric spheres and had the

Eudoxus' theory of Mars

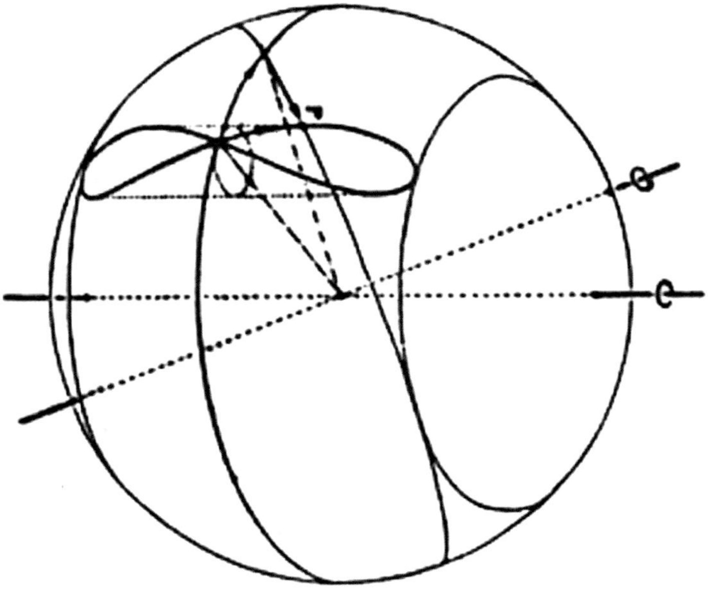

advantage that all the motions centered on the Earth. It was certainly ingenious, and it worked – up to a point. Unfortunately, it rather too generously made Mars retrograde three times instead of only once as observed. Even more fatally it could offer no explanation of why the planet varied through such a wide range in brightness. (For that matter, why *should* a planet's brightness vary at all in an Earth-centered scheme?)

We do not know how Eudoxos himself regarded his invention; perhaps he regarded it as nothing more than a mathematical recreation. Moreover, there is not the slightest reason to suppose that he believed in the literal existence of his spheres, though later on – especially in the writings of Aristotle – their physical reality was claimed and rather robustly asserted. A century after Eudoxos, the mathematician Archimedes amused himself by producing a model of the Eudoxan system using glass spheres turned by water power. But by then Eudoxos' model was nothing more than a toy.

All the Greek astronomers had abandoned it, and turned to other planetary models in which – as dictated by the observed variation in the brightness of the planets – they still moved in circles, but the center of each circle was slightly offset from, or eccentric to, the Earth.

They called this device the "movable eccentric," and it had the advantage of bringing the planet closer in at one point in its circuit and moving it further away at another. Thus it could account for why Mars would appear bright at one time and dim at another. Furthermore, since Mars did not always come to opposition at the same point of the zodiac, the circle in which the planet moved was made to pivot slowly around the Earth.

The movable eccentric did an admirable job of producing the retrograde movements of the outer planets, and it worked just as well for the inner planets, Mercury and Venus. In the latter case, the centers of the planetary circles were always required to remain on a straight line between the Earth to the Sun so as to keep them confined within a limited angular distance of the Sun.

Heracleides of Pontus (c. 390–after 322 BC), like Eudoxos another of Plato's pupils, proposed that the apparent diurnal rotation of the heavens was due to a west-to-east rotation of the Earth's sphere every 24 hours and added a further simplification by centering the circles of Mercury and Venus on the Sun instead of merely keeping them lined up with it.[29] His scheme – which became known as the "Egyptian system" – was already halfway to a true heliocentric model. But the Greeks were never ones for half-measures, and the final step was taken soon afterwards by the next great Greek astronomer, Aristarchos of Samos (310–230 BC).

Aristarchos was active by at least 281–280 BC when, as Ptolemy records, he observed the summer solstice. Only one of his works has survived, "On the distances of the Sun and Moon," in which he describes an ingenious method of using the Sun–Earth–Moon angle at the time the Moon's phase is exactly half to work out the relative distances of the Sun and the Moon. Aristarchos uses a figure of 87° for the Sun–Earth–Moon angle, which proves that he can hardly have tried to measure anything (how he came to adopt this figure is a complete mystery). Even with rather laughably modest means, such as using the post of a backyard clothesline as a gnomon, my friend Art Hoag was able to show that this angle had to be at least 89°; more accurate measures show the true figure to be 89.8° from which it follows, using Aristarchos's own method, that the Sun must be almost 400 times farther from the Earth than the Moon.[30] But even the 87° figure, crude as it is, implies that the Sun must be many times farther from the Earth than the Moon, and hence much larger than the Moon or even the Earth. (This was certainly progress; by contrast, Anaxagoras of Clazemenae, a close friend of Pericles, in the fifth century BC, had scandalized the religious conservatives among the Athenian public by arguing that the Sun was a red-hot stone larger than the Greek peninsula of Peloponnese. Yes, it is – *rather*!).

If, then, the Sun is so much larger and more noble then the Earth, might it not make sense for the Earth to travel around the Sun rather than the other way around? We can well imagine that Aristarchos might have thought so, but unfortunately we

---

[29] What is reported of Heracleides is not very flattering. Bertrand Russell says, in *History of Western Philosophy*, p. 223: "he must have been a great man, but was not as much respected as one would expect; he is described as a fat dandy."

[30] Arthur A. Hoag, "Aristarchos Revisited," *Griffith Observer*, 54 (1990):10–18.

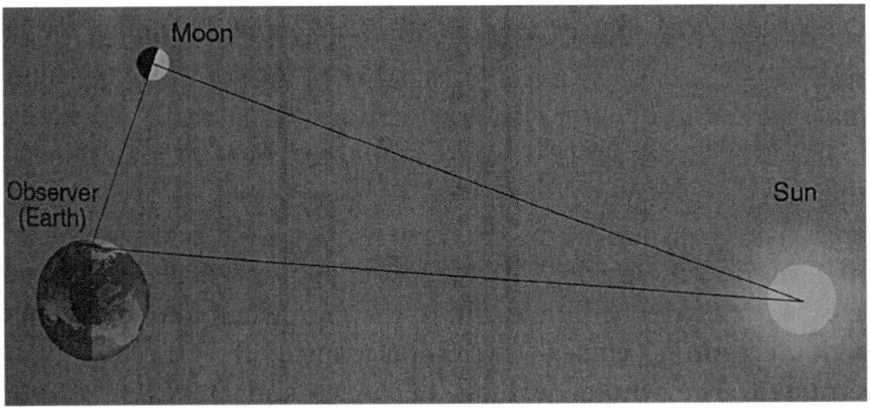

Aristarchus's scheme for determining the distance to the Moon

do not know. Somehow – before the middle of the third century BC – he came to that conclusion and asserted that in addition to the rotation on its axis it performed in Heraclides' scheme, the Earth also made an annual circuit around the Sun. Perhaps Aristarchos realized – just as Copernicus would realize 1,700 years later – that the retrograde motions of Mars and the other planets could be elegantly derived from such a model. As a feat of sheer mathematical virtuosity, Aristarchos' heliocentric system is one of the most brilliant achievements of antiquity. Did he himself take it seriously; or was he only amusing himself by drawing figures in the sand?

What we do know is that the idea was admired, despised, debated, and in the end rejected by his contemporaries. With the sole exception of Seleucus of Seleucia, who espoused it more than a century later, Aristarchos' system found no known followers in the ancient world. To the orthodox, his idea was regarded as outright dangerous. Things came to a head when Aristarchos was living in Athens. The religious conservatives of the city were just as scandalized by his striking hypothesis as they had been by Socrates and other philosophers who called into question traditional beliefs. It is not always appreciated that, as E. R. Dodds points out, "the Great Age of Enlightenment" in Greece was also "like our own time, an Age of Persecution."[31] Plutarch observed in the second century AD:

---

[31] E. R. Dodds, *The Greeks and the Irrational*. Berkeley: University of California Press, 1951, p. 189. Anaxagoras had been prosecuted because he dared reduce the celestial divinities into stones and earth. He escaped the death sentence by choosing exile. Socrates was prosecuted "...for not believing in the gods of the city-state, but in other new divinities." We learn from Plato's *Apology* that at his trial Socrates refused to be associated with the astronomers, and disclaimed any knowledge of astronomy; in contrast to the astronomers, of whom he says "those who hear them think that men who investigate these matters do not even believe in gods." Socrates claimed to the contrary that he does "... believe that the sun and moon are gods, like all the other people

> There was widespread intolerance of natural scientists and "star-gazers," as they were called at the time, on the grounds that they reduced the divine to irrational causes, blind forces and necessary incidents. Hence it was that Protagoras was banished and Anaxagoras cast in prison and rescued with difficulty by Pericles, and Socrates, though he had nothing whatever to do with such matters, nevertheless lost his life, because of philosophy.[32]

Effectively, Aristarchos had proposed that the definition of "planet," accepted from time immemorial to include the Moon, Mercury, Venus, the Sun, Mars, Jupiter, and Saturn, be revised. The Sun was out, the Earth was in. But he was voted down, "plutoed," by whatever the equivalent of the International Astronomical Union was at the time. Religious prejudice no doubt played a role; it appears that in Athens Aristarchos was accused of impiety by Cleanthes, the head of one of the schools at Athens (the Stoic school). According to Plutarch again:

> Cleanthes thought that the Greeks ought to lay an action for impiety against Aristarchos the Samian on the ground that he was disturbing the Hearth of the Universe [the Earth], because he sought to save the appearances by assuming that the heaven is at rest while the earth is revolving along the ecliptic and at the same time is rotating about its own axis.[33]

If he had remained in Athens, Aristarchos might have faced the death penalty; he had no wish to be another Socrates, and left Athens, never to return. In addition to the usual opposition of the troglodytes, Aristarchos' idea was also rejected by other philosophers for whom it seemed to fly in the face of common sense. It seemed evident to them that, as E. A. Burtt has written,

> The Earth was a solid, immovable substance, while the light ether and the bits of starry flame at its not too distant limit floated easily about it day by day. The Earth is to the senses the massive, stable thing; the heavens are by comparison, as revealed in every passing breeze and flickering fire, the tenuous, the unresisting, the mobile thing.[34]

Aristarchos' idea also made the universe much larger than anyone at the time was prepared to believe. Archimedes (c. 290–280 to 212/211 BC), the brilliant Greek mathematician of the generation after Aristarchos, explained to his patron King

do." He was prosecuted anyway, and chose to drink the hemlock rather than accept exile. Xenophon (*Memorabilia*, IV, 7, 6f) provides even more detail about Socrates' views regarding astronomy. "With regard to the phenomena of the heavens, he disapproved strongly of attempts to work out the machinery by which the god operates them; he believed that their secrets could not be discovered by man, and that any attempt to search out what the gods had not chosen to reveal must be displeasing to them. He said that he who meddles with these matters runs the risk of losing his sanity as completely as Anaxagoras, who took an insane pride in his explanation of the divine machinery... When [Anaxagoras] pronounced the sun to be a red-hot stone, he ignored the fact that a stone in fire neither glows nor lasts long, whereas the sun-god shines with unequalled brilliance forever."

[32] Plutarch, "Nicias," 23, 2–3. Quoted in: Ioannis Liritzis and Alexandra Coucouzeli, "Ancient Greek helicoentric views hidden from prevailing beliefs," *Journal of Astronomical History and Heritage*, 11, 1 (2008), 39–49:43.

[33] Plutarch, "On the Face in the Orb of the Moon," 6, 923a.

[34] E. A. Burtt, *The Metaphysical Foundations of Modern Science*. Garden City, NY: Doubleday, 1954, p. 37.

Gelon of Syracuse that Aristarchos seemed to believe "the sphere of the fixed stars, situated about the same center as the Sun, is so great that the circle in which he supposes the Earth to revolve bears such a proportion to the distance of the fixed stars as the center of the sphere bears to its surface." That, however, seemed inconceivable to him. It was a conclusion whose very outlandishness seemed to serve as its own refutation – a case of "*reductio ad absurdum*."

Though we lack the details, it seems that in the next generation after Aristarchos, Greek astronomers began to gain access to the highly accurate records that had been painstakingly accumulated over centuries by the Babylonians. A few not only realized the potential usefulness of this data but the importance of making their own observations. The significance of this cannot be too emphatically stated.

The Babylonians, as we have seen, had been meticulous record-keepers of the times of events such as the heliacal risings of planets and eclipses for centuries, but never attempted to derive general laws from their data. The Greeks, on the other hand, had begun with sweeping generalizations with very little data to support them – "all is water," "all is air," "all is number," and eventually, as we saw, they got to the point where they concerned themselves largely with mathematical abstractions – objects that existed only in a pure Platonic heaven and could be apprehended only in terms of pure thought.

Someone by the end of the third century BC – the generation or so after Aristarchos – saw the need to move back from the abstract to the concrete, from the qualitative models of planetary motion to the quantitative data of measurement. In other words, it appeared that the geometry of the Greeks had to be combined with the rigorous quantitative data of the Babylonians. This fusion led to the first true models of the planetary motions – models that could be rigorously tested by observations, were constrained by numerical data, and could be modified and perfected to improve their agreement with the phenomena they were meant to represent.

It appears that the decisive marriage of Greek and Babylonian thought was made possible by the conquests of Alexander the Great (who died, in fact, in Babylon, which he had made part of his vast empire, on his rather straggling and drunken way back from the Indus, in 323 BC). Acquaintance with Babylonian methods stimulated the Greeks of the Hellenistic period – the era that followed Alexander's startling career, when his conquests were trisected among his three generals – to become observers of as well as speculative thinkers about the stars. Thus, as noted by astronomical historian Antonie Pannekoek, "The Babylonian results for the periods and irregularities which had remained simply as numerical data, became in the hands of the Greeks the basis of geometrical constructions and led to the conceptions of spatial world structure."[35]

Possessed of the Babylonian numerical treasure trove, the Greeks of this period seem to have assimilated the synodic periods of the planets (the times needed for them to return to the same positions relative to the Sun). They knew the lengths and

[35] Pannekoek, *History of Astronomy*, p. 123.

speeds with which the planets traversed their retrograde arcs and, in the case of Mars in particular, their variations from opposition to opposition. The recognition of these "second-order" irregularities in the motions (the variation in the lengths and speeds of the retrograde arcs) showed them that the heavens had to be tinkered with again.

But if further tinkering were needed, these later astronomers preferred to work with their observing platform assumed to be the stable and stationary Earth, securely anchored in place, rather than following Aristarchos by adopting the reckless expedient of sending it hurtling wildly around the Sun.

The tinkering began in earnest with the work of Apollonius of Perga (c. 262–c. 200 BC). As a young man, Apollonius left Perga (in Asia Minor) for Alexandria, attracted, we are told, by Aristarchos' fame. His mathematical abilities are attested by the quality of his great work on the conic sections, and he developed a theory of the planetary motions in which he wielded with considerable virtuosity the mathematical device of the movable eccentric as well as an ingenious new device, the epicycle, which he probably invented.

Here, in a nutshell, is how the epicycle worked. In a geocentric system, the Sun moves around the Earth with a period of 1 year. The other planets move around the center of a little circle in the direction shown, which in turn pivots around a larger circle centered upon the Earth. The little circle is the epicycle, the larger one the deferent.

Retrograde movement produced by the device of the epicycle

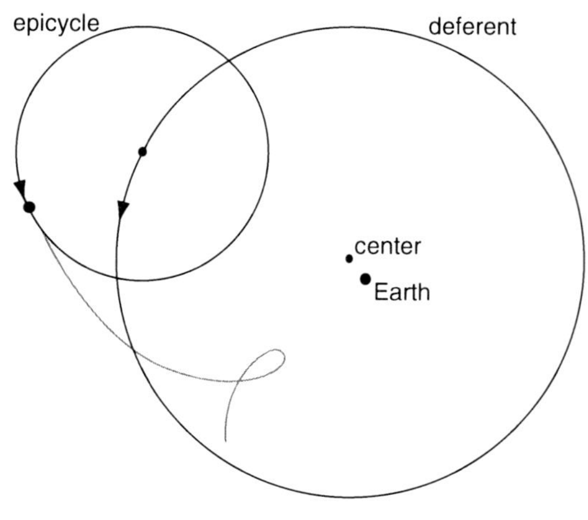

Apollonius himself seems to have worked out that the movable eccentric and the epicycle are mathematically equivalent. That being the case, they can be used interchangeably. Since the epicycle construction is more intuitive and easier to work with, it soon came into more or less exclusive use. It is essentially a gear-mechanism of the kind used in clockwork. For a long time it was assumed that because the

Epicycles of Jupiter

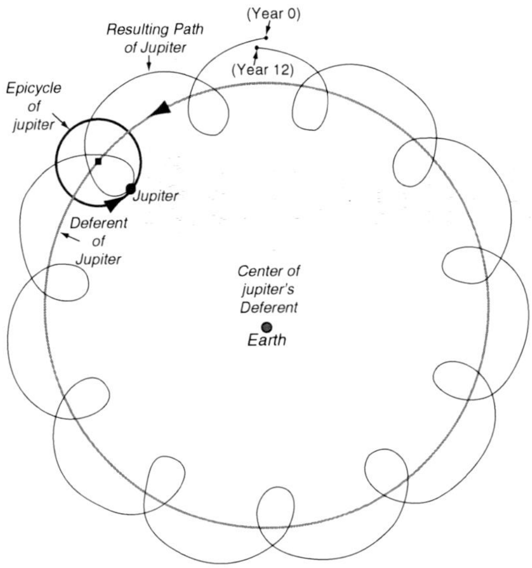

Greek philosophers asserted that reality was grasped most directly by the intellect alone and not from the senses – and because many of them viewed with contempt artisans and manual laborers – their astronomy was developed largely by strictly geometrical means like those of Euclid. That was certainly true of earlier Greek astronomers. But we now know that the Greek astronomers of the third and second centuries also were capable of constructing sophisticated mechanical devices, whether as representations of the beautiful clockwork universe they had worked out – "something to elevate the spirit and get closer to God or the true meaning of things," as Jo Marchant puts it[36] – or to assist them in their calculations. (As earlier noted, Archimedes himself had become rather famous for such mechanical inventions, though Plutarch, with his disdain of all manual labor, insisted that he could only have regarded them as "sordid and ignoble" toys.)

Apollonius' greatest successor was Hipparcos of Rhodes, who lived in the second century BC. He must have been a mechanical genius and clever with his hands, since he possessed instruments he can only have designed himself and with which he made observations so accurate that they remained unsurpassed until the time of Tycho. Unfortunately, we have no idea what those instruments were. Among his results, he determined the duration of the year to within 6 min, calculated the inclination of the Earth's axis (his value, 23°51′, compares favorably to the correct value at the time, 23°46′), and discovered the evection of the Moon (the perturbation of the Moon's

[36]Jo Marchant, lecture to the Royal Institution in London, quoted in: "Antikythera clockwork computer may be even older than thought," *Guardian science blog*, July 29, 2009.

motion due to the attraction of the Sun). But his most remarkable discovery was the precession of the equinoxes, which he worked out by comparing modern observations of eclipses with old Babylonian records. This is the slow shift in the position of the equinoxes due to the wobble of the Earth's axis produced by the pull of the Moon.

Perhaps his mechanical skill is even attested in the remarkable Antikythera mechanism, found in a wrecked ship discovered in 1901 by sponge fishermen off the coast of Antikythera, between Kythera and Crete.[37] The wreck contains vases in the style of Rhodes, suggesting the ship was en route from Rhodes when it sunk. The association with Hipparcos' Rhodes is suggestive as is the probable date of the wreck – the second half of the second century BC. The Antikythera mechanism is a clockwork that consists of at least 30 interlocking gear-wheels used to compute and display the motions of the Sun and the Moon accurately enough to predict the dates of their eclipses, and possibly also the motions of the planets.

Apart from the circumstantial evidence tying the device to Hipparcos – the ship's origination in Rhodes and sinking not long after Hipparcos is known to have been active there – the most convincing clue is that the antikythera mechanism demonstrates the intricate motion of the Moon on the basis of a theory Hipparcos himself worked out. One wheel moves around once every 9 years, the period in which the Moon's perigee – the point of its orbit closest to the Earth – completes one complete swing around the Earth. To this wheel are fixed a pair of small wheels, one almost centered on top of the other; the bottom wheel has a pin sticking up from it which pushes the top wheel around. Because the two wheels are not exactly centered, the pin moves back and forth and causes the movement of the upper wheel to speed up and slow down, as the Moon actually does.[38]

Though we have only this one remarkable mechanism, a chance survivor because it ended up on the bottom of the sea, others like it must once have existed. Unfortunately, being made of bronze, they were likely to be melted down for scrap when no longer needed. Such mechanisms may have been more important than we realize in motivating the development of Greek astronomical theory in the third and second centuries BC, when clearly the Greeks were not so entirely mentalist in their approach as might be imagined from the writings of people like Plato or Plutarch.

Since their original works on astronomy have not survived, what we know of Apollonius' and Hipparcos' contributions we owe to the great summing up by Claudius Ptolemy, who lived at Alexandria, Egypt, in the second century AD.

---

[37]An excellent account is: Jo Marchant, *Decoding the Heavens: the mystery of the world's first computer*. London: William Heinemann, 2009.

[38]See: M.T. Wright. Epicyclic gearing and the Antikythera Mechanism, Part I. *Antiquarian Horology*, 27 (2003), 270–279 and Part II, 29 (2005), 52–63. Also: M.T. Wright. The Antikythera Mechanism and the early history of the moon-phase display. *Antiquarian Horology*, 29 (2006), 319–329. The Antikythera mechanism was used to calculate the 4-year cycle of the Olympiads and its associated pan-Hellenic games and contains a dial, arranged as a five-turn spiral, that is a 19-year calendar based on the Metonic cycle, while the lower dial is a Saros eclipse-prediction dial, arranged as a four-turn spiral of 223 lunar months, with glyphs indicating eclipse predictions. See: T. Freeth, A. Jones, J.M. Steele and Y. Bitsakis, Calendars with Olympiad display and eclipse prediction on the Antikythera Mechanism. *Nature*, 454, 7204 (July 31, 2008), 614–617.

Ptolemy. This drawing, inspired by the fifteenth century sculptor Jörg Syrlin the Elder's figure on the choir stalls at the Cathedral at Ulm, was made by Randall Rosenfeld. © Randall Rosenfeld

The Pantheon, rebuilt by the emperor Hadrian during the time of Ptolemy. Photograph by William Sheehan of a painting in the Vatican Museum, Rome

Interior of Pantheon. Photograph by William Sheehan, 2008

Egypt was then a province of the far-flung Roman empire of the Antonines. During Ptolemy's lifetime the Pantheon, the Temple to all the Planetary Gods, was finished under Hadrian, and boasted antiquity's most noble dome (still to this day the world's largest unreinforced concrete dome). Hadrian was succeeded by Antoninus Pius and Marcus Aurelius, who are remembered as the best and wisest of the Roman emperors. The historian Edward Gibbon, in a famous passage of *The Decline and Fall of the Roman Empire*, called this era unquestionably the period of the world "during which the condition of the human race was most happy and prosperous," and Ptolemy was the greatest astronomer of this ostensibly golden age.

He was an encyclopaedist, employed on colossal projects, concerned with vast things. He was working on a five-foot shelf of treatises on every aspect of applied mathematics: harmony, optics, geography. So wide-ranging were his interests that he even wrote a book on astrology (*Tetrabiblos*), and was obviously quite sincere about it; he seems to have regarded it as a branch of "applied mathematics." One may wonder that the same hand produced both the *Tetrabiblos* and the *Almagest*, but – as Owen Gingerich has pointed out – in the second century AD astrologers would have been the major consumers of planetary theory![39]

In the Middle Ages, Ptolemy's reputation was largely founded on his *Geography*, a work now largely forgotten. Naturally, he made mistakes; his most serious was to follow Strabo instead of Eratosthenes in his estimate of the length of a degree

[39]Owen Gingerich to William Sheehan; personal communication, December 26, 2006.

(80 km/50 miles instead of 113 km/70 miles), which led him to greatly underestimate the circumference of the Earth. Coupled with his mistake of extending Asia far beyond its true dimensions, this had the effect of grossly reducing the extent of the unknown parts of the world between the eastern tip of Asia and the western tip of Europe. But even Ptolemy's mistakes were often fruitful: this one had the effect of hastening the European encounter with the New World.

Ptolemy's greatest astronomical work is the μεγ στη [*Syntaxis*] (later known as the Great Syntaxis to distinguish it from a lesser collection of astronomical writings; the title was later rendered by the Arabs as *al-majisti*, the Greatest, and translated from Arabic to Latin as the *Almagest*, by which it is generally known). It seems to be, in date of composition, the earliest of Ptolemy's works. Next to Euclid's *Elements*, it has the distinction of having been the scientific text longest in use, and remained the supreme authority on astronomy wherever Greek culture and learning survived for over a thousand years.

Ptolemy was an assimilator and perfecter of the existing planetary theories and so successful that the technical details of the works of his predecessors have not survived. He was the last of a distinguished line of Greek technicians, a consummate craftsman of astronomy based on the circle. Ptolemy brought the Earth-centered model of planetary motion to near-perfection but not, as is sometimes alleged, by simply piling circles on circles until he had produced an unwieldy, artificial "crank machine." Like poetry, his system does not withstand paraphrase. I agree wholeheartedly with Owen Gingerich's assessment: "It is difficult to convey the elegance of Ptolemy's achievement to anyone who has not examined its details. Basically, for the first time in history (so far as we know) an astronomer has shown how to convert specific numerical data into the parameters of planetary models, and from the models has constructed a … set of tables … that employ some admirably clever mathematical simplifications, and from which solar, lunar, and planetary positions and eclipses can be calculated as a function of any given time."[40] Similarly Albert Einstein claimed that "among his astronomical predecessors he had the greatest admiration for Ptolemy, because of the fact that although his instruments were of the most primitive he got very close to a number of modern values for astronomical quantities."[41]

Ptolemy could take special satisfaction in his theory for Mars. It was indeed ingenious, and turned on what must be regarded as his master stroke. He effectively solved a problem that had defeated Hipparcos as well as Apollonius and Eudoxos before him and which, by 135 AD, could be counted as the outstanding unsolved problem of astronomy and had earned Mars its reputation as the untrackable and

[40] Owen Gingerich, *The Eye of Heaven*. New York: American Institute of Physics, 1993, p. 55.

[41] In a conversation with Janos Plesch, a Hungarian doctor who had been Einstein's personal physician in Berlin until 1933 and maintained a friendship with him over the years, a few days before Einstein's death. Quoted in: Janos Plesch and Peter H. Plesch, "Some reminiscences of Albert Einstein," *Notes and Records of the Royal Society of London*, 49:2 (1995), 303–328:317.

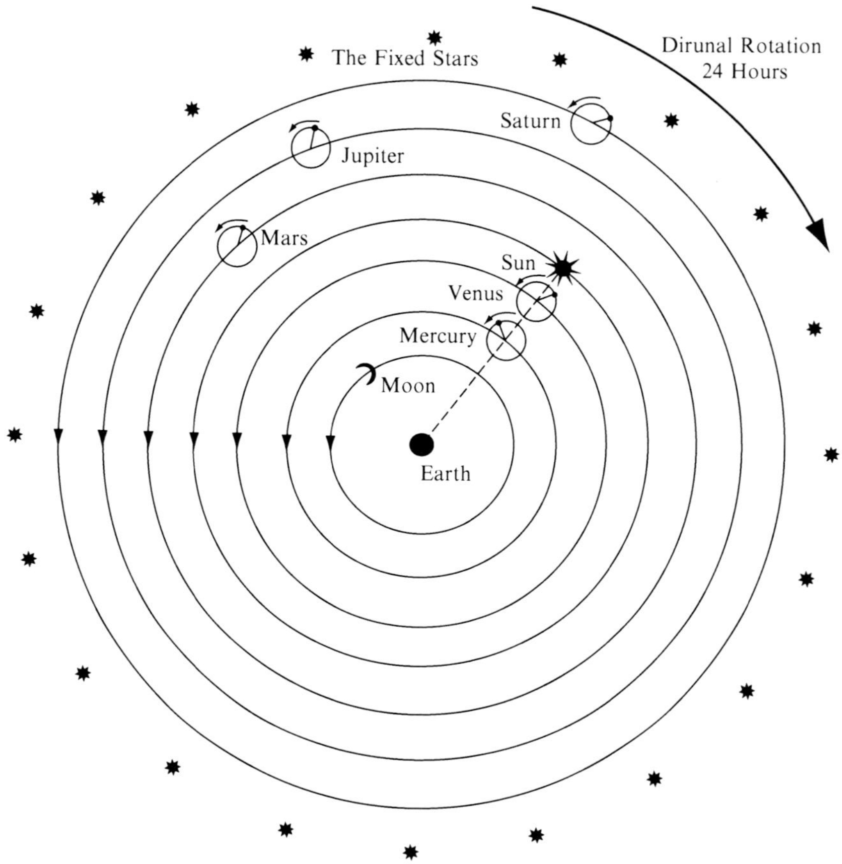

Ptolemaic system

intractable "star." What Ptolemy inherited from his predecessors was the basic epicycle construction with which we are already familiar – by the same token, Bach inherited the fugue and Beethoven the sonata. The way he applied it was ingenious.

As the epicycle turned, a point on its rim followed a looped path swinging in toward the center of the deferent, before moving outward again in reverse, just as was called for to explain the retrograde motion of the planet. Though this basic appliance had worked well enough in Apollonius' day, it was no longer acceptable – at least not without modification – in Ptolemy's. Ptolemy, after all, possessed much more accurate observations than Apollonius had possessed, and knew that Mars's motion around its orbit is markedly nonuniform. When it comes to opposition in Capricorn/Aquarius, Mars is moving twice as fast as when on the other side of its orbit in Cancer/Leo. In order to account for this variation, Ptolemy positioned his deferent circle slightly off-center from the Earth until he had the speed of Mars's motion just right. With the Earth off center of the Martian orbit, he could also account for the planet's marked variations in brightness. Unfortunately, there was still a problem – the size of the retrograde loops did not match what was observed. Here Ptolemy introduced his brilliant expedient. He made Mars move uniformly not around the center of the deferent but around another point which he called the *equant*, as it was positioned an equal distance on the opposite side of the center from the Earth (this solution was referred to as the "bisection of the eccentricity"). Now the whole scheme worked beautifully, though at the expense of breaching the rule of uniform circular motions that had been dogmatically insisted upon by all astronomers since the time of Plato.

Ptolemy's theory made the radius of Mars's epicycle parallel to that of the Earth–Sun radius. By means of this arrangement, he ensured that the retrograde loop of Mars always occurred whenever Mars was at opposition. Gratified at his success with Mars, Ptolemy applied it without hesitation to the rest of the planets – with one exception. Mercury, the innermost planet to the Sun, has always been the most difficult to observe. Ptolemy, making the best of a bad situation and too deferential it would seem to data of doubtful value, gave it a complicated path with a double perigee (two points, one in February and the other in July, at which Mercury was made to swing inward toward the Earth). Here at least Ptolemy seems to have deserved the epithet King Alfonso X of Castile later flung at the whole edifice. Alfonso memorably referred to it as "a crank machine." In fact, Ptolemy's theory of Mercury incorporated a device that worked exactly like a crank – the center of the deferent moved about a little circle that thrust the epicycle back and forth.

Parenthetically, Ptolemy's device would reappear at the end of the eighteenth century when the English inventor James Watt borrowed what he called the "Sun-and-planet gear" in his designs for the steam engine. It is effectively the same crank-mechanism that Ptolemy had devised for Mercury. So Ptolemy – often unfairly denigrated as a fusty conservative and an uninspired jobber – turns out to have been an unheralded forerunner of the Industrial Revolution.

Ptolemy's system is a complicated, intricate, and elegant mechanism. With hindsight – and indoctrinated in the heliocentric point of view – we can easily enough see that the main epicycles for each planet are reflections of the Earth's motion around the Sun, and that a simplification results by putting the Sun in place of the Earth. How did Ptolemy not see this, we are tempted to ask, and in so doing put astronomy in a more productive path for the next 15 centuries?

But it is unfair for us to ask, and we must keep ourselves to what was possible to the human intellect in the second century AD. The motions of the planets must then have seemed a problem as intractably difficult as superstring theory does to us today. At a conference on "the Unity of Mathematics" at Harvard in September 2003, the mathematician Sir Michael Atiyah gave an interesting talk on "the interaction between geometry and physics," at which he said:

> If we end up with a coherent and consistent unified theory of the universe, involving extremely complicated mathematics, do we believe that this represents "reality"? Do we

Sun and planet gears. The fact that Ptolemy's theory of the motion of Mercury anticipated the Sun and planet gear mechanism of James Watt's steam-engine is the inspiration of this painting by Julian Baum. © Julian Baum

> believe that the laws of nature are laid down using the elaborate algebraic machinery that is now emerging in string theory? Or is it possible that nature's laws are much deeper, simple yet subtle, and that the mathematical description we use is simply the best we can do with the tools we have? In other words, perhaps we have not yet found the right language or framework to see the ultimate simplicity of nature.[42]

Ptolemy – with the tools at his disposal – was successful enough to show what had by no means been obvious, say, in the time of Eudoxos: that the planetary motions were rational, and ultimately within reach of solution by the human intellect. It is something we cannot yet say for sure regarding the whole project of superstring theory today. The current difficulties of superstring theorists in finding a simple equation, beautiful idea, or fundamentally symmetry principle that will explain the intricate structures they have been studying causes us to view with sympathy, and even envy, the lot of old Ptolemy. Perhaps his epicycles do seem to us artificial and inelegant; but they are no more so than "the current best picture of the world provided by actual existing superstring theory [which] is neither beautiful nor elegant. The 10- and 11-dimensional supersymmetric theories [currently being used which] are very complicated to write down precisely. The six- or seven-dimensional compactifications of these theories necessary to try to make them look like the real world are both exceedingly complex and exceedingly ugly."[43] I suspect that if Ptolemy were living today, he would be a superstring theorist – and one of the best of them.

[42] Quoted in Peter Woit, *Not even Wrong: the failure of string theory and the search for unity in physical law*. New York: Basic Books, 2006, p. 262.

[43] Woit, *Not even Wrong*, p. 262.

There is an evocative wooden sculpture of Ptolemy in the Cathedral of Ulm: eyes closed, a faint smile on his face, he holds an armillary sphere in one hand while pointing casually upward toward the heavens with the fingers of the other. It is an image of smugness and complacency – the face of an astronomer so satisfied with the completeness of his theory that he no longer bothers to look into the heavens (and certainly he was not primarily an observer; he borrowed most of his observations from his predecessors – especially Hipparcos). But Ptolemy cannot have remotely resembled this image. Instead he must have been one of the most passionate seekers after celestial truth that has ever lived. We even seem to find an expression of the ecstasy that his models of the heavens brought to him a marvelous passage found in some of the old manuscripts of the *Almagest* which I would like to believe was written by the great man himself:

> I know that I am mortal by nature, the creature of a day;
> But when I trace at my pleasure the windings to and fro of the heavenly bodies
> I no longer touch the earth with my feet:
> I stand in the presence of Zeus himself
> And take my fill of ambrosia, the food of the gods.[44]

Ptolemy's last recorded observation was made in 151 AD, and he must have lived on to about 180. His planetary system is the epitome of what the Greek mind had accomplished in astronomy. We have come a long way from the crocodile-men of Mycenae, or even the number-mysticism of Pythagoras.

We may not always appreciate, by the way, just how far Ptolemy was in time from the first Ionian philosophers, like Thales and Pythagoras. He was as far remote from them as we are, say, from William Caxton and Geoffrey Chaucer.

By the time Ptolemy died, happiness had largely departed from the ancient world. Despite Marcus Aurelius's own admirable personal qualities and his stoic fortitude, he presided over a realm beset by a series of unmitigated calamities. (He died in 180 AD, quite possibly the very same year as Ptolemy.) Though Gibbon described the Roman Empire under the Antonines as one of undisturbed peace and prosperity, in fact the society was "rotten to the core." Pannekoek observes:

> The sunset glow of antiquity shed its light radiance upon a worn-out world. Before long, towards the end of the second century, the storms broke loose that within the space of a single century were to undermine the power of a world empire and, after another century, were to lay waste the foundations of the ancient world and its culture.[45]

We have seen this – systemic failure – before. It occurred at the end of the Bronze Age in the eleventh century BC, when the Mycenean world disintegrated. The causes of the ultimate collapse of the classical world has long been debated; among the indisputable ones is the fact that the labor system of the ancient world from the beginning had been based on slavery, and with the end of the wars of conquest, the supply of slaves was checked. Moreover, the gold and silver mines

[44] Quoted in Gingerich, *The Eye of Heaven*, p. 4.

[45] Pannekoek, *History of Astronomy*, p. 161.

were exhausted, and the economy devolved to primitive agriculture. The entire Italian peninsula depended on grain from remote provinces and for protection on mercenary armies. A pestilence broke out in 188 AD, brought back by armies from Asia. Adding aggravation to aggravation were floods and famines in Italy, earthquakes in Asia. The barbarian tribes began to press in and settled in the depopulated areas of the contracting empire. Even Marcus Aurelius himself had spent much of his time on the empire's frontiers fighting defensive wars against the Parthians in the east and the Germans in the north; during the next century a series of short-lived successors – appointed and then murdered at the pleasure of the army – watched with increasing alarm the encroaching dangers, and raced to build defensive works from the Veneto to Milan. Rome itself was surrounded with massive walls and fortifications. Instead of marking the farthest limits of the imagination in its indomitable struggle with the unknown, the word "frontier" then acquired the meaning it still has in Europe – the sharp edge of sovereignty, a line to stop at not an area inviting entrance. It was the outermost wall of fortifications on a verge of fear.

Inevitably, shells of dogma hardened to encase the human mind. Greek science, whose seed had first been planted in Ionia in the sixth century BC, had been strengthened in the hybrid it formed with Babylonian science and, after branching and prospering for centuries, had petered out into a magnificent but apparently final twig in Ptolemy's complicated orrery. Then, like a species bound for extinction or a flickering candle blown out by the wind, the impulse to understand the heavens – like a passionate love-affair abruptly ended – seemed to have gone dead. The seed had dropped back into the earth, and would remain in a state of dormancy for more than a thousand years.

Chapter 6
Revolutions

By what authority does Man assume that this admirable moving of Heaven's vault, the eternal light of these lamps burning so proudly over his head ... were established and continued so many ages for his commodity or service?

Michel de Montaigne

Johannes Vermeer's Astronomer in the Louvre. Photograph by William Sheehan, 2009

W. Sheehan, *A Passion for the Planets: Envisioning Other Worlds, From the Pleistocene to the Age of the Telescope*, DOI 10.1007/978-1-4419-5971-3_6,

By the time Ptolemy died, in the last quarter of the second century AD, classical civilization was already in decline, and would soon collapse. Meanwhile, Christianity had taken root. Though the classical and Christian traditions overlapped chronologically and geographically, they were utterly different in spirit and inspiration. Christianity was largely derived from Eastern mystery religions and the Hebrew Bible – the latter, however, drastically reordered and "misread," as Harold Bloom points out, in getting to the New Testament.[1]

At first a despised cult, the Christian religion was placed under the protection of the Roman empire by Constantine's edict of Milan in 313 AD, and thereby imposed the autocratic style of Rome on the church as much as it Christianized the Roman empire.[2] In the following century, confederations of Germanic tribes that had begun to gather on the Rhine and the Danube after Marcus Aurelius's death surged forward – pushed, it seems, by migrating tribes of the Huns who had themselves originated in the remote Mongol Steppe.

In 410, Rome itself was sacked for the first time in 800 years, by the Visigoths under Alaric. Within a few decades, the Visigoths had carved out a kingdom in southern Gaul and Spain, while another tribe, the Vandals, had taken control of North Africa. Even so, the lapse of Roman rule in the West did not yet mark the "Fall of the Roman Empire," for the Germanic tribes also adopted Christianity. The vandalized empire's heir and representative became the Roman Catholic Church.

Among the Greeks of classical times, there had always been a keen love of debate; "a commoner," writes Richard Nisbett, "could challenge even a king and not only live to tell the tale, but occasionally sway an audience to his side... Uniquely among ancient civilizations, great matters of state, as well as the most ordinary questions, were often decided by public, rhetorical combat rather than by authoritarian fiat."[3] The Christian relationship to language was entirely different, however; the Scriptural texts – those established as "canonical" at any rate – were authoritative; different views were settled by councils – or by fiat – after which they were not open to further question or debate. Any contrary position was deemed heretical, and to hold to it potentially put one's soul in mortal danger. Religious dogma is, by its very nature, ideological and inimical to empiricism. It demands unquestioning faith and blind obedience.

What wonder that during the long period of European history that Will Durant describes as the "Age of Faith," science – and honest inquiry – suffered a near-death and commenced upon a Rip van Winkle-like sleep of a thousand years. The ardor of blind faith burned brightly during those centuries, and – as "intolerance is the natural concomitant of strong faith; tolerance grows only when faith loses certainty; certainty is murderous"[4] – the literal fires burned any and all deemed troublemakers.

---

[1] Harold Bloom, *Jesus and Yahweh: the names divine*. New York: Riverhead Books, 2005, pp. 56–47.

[2] See Cornel West, *Democracy Matters: winning the fight against imperialism*. New York: Penguin, 2004, pp. 146–149.

[3] Richard Nisbitt, *The Geography of Thought: how Asians and Westerners think differently ... and why*. New York: Free Press, 2003, p. 3.

[4] Will Durant, *The Age of Faith*. New York: Simon and Schuster, 1950, p. 784.

Needless to say, there was little interest in the planets during these centuries of inconclusive theological controversy. Alexandria, the city founded by Alexander the Great in the fourth century BC, the center of Greek mathematics and science from the time of Eratosthenes until that of Ptolemy, and the place where the miscegenation of Greek geometry and Babylonian numerical data about the planets had occurred, now became a center of Christian orthodoxy. The great library was destroyed by Christian fanatics in the third century AD, then again three centuries later by a Muslim army under Caliph Omar, called the "Emperor of Believers;" who died in 644 AD. The ancient learning became generally scattered. The world of these centuries was not only economically depressed but intellectually impoverished compared to the brilliant Hellenistic civilization which had preceded it. Though it has become unfashionable to speak of well-demarcated "periods" like the "Dark Ages" or the "Enlightenment," these centuries afford little to command our attention here. Much of the credit for keeping whatever was left of the ancient learning alive fell to the later Arabs who did not, fortunately, follow the example of Caliph Omar, and whose important contributions are only now being reevaluated in the light of modern scholarship.[5]

Arab astrolabe in the Oxford Museum of the History of Science. Photograph by William Sheehan, 2009

[5]George Saliba, *Islamic Science and the Making of the European Renaissance*. Cambridge, MA: MIT Press, 2007.

In Christian Europe during the Middle Ages, there was a retrogression. Space consisted of the flat Earth, with the heavens regarded as an inverted bowl in which the stars were peepholes into the flaming firmament beyond. The Moon, Sun, and planets moved in epicycles whose dimensions, though indeterminate, were encompassed by the crystalline spheres. Even hell still existed as a geographical location within the Earth, and was more fearfully real to Medieval men and women than we today can possibly imagine.

Men and women of that time believed in a narrowly circumscribed Ptolemaic universe that resembled nothing so much as a Medieval walled city. Likewise, the sense of time was no less circumscribed; it was largely fixated on the past – on the events of the life of Christ – and so there was no real interest in time-keeping except in maintaining the tables used to determine the dates of Easter from which followed all the other dates of the moveable feasts of the Christian calendar. William Manchester has summed up the mentality of that vast interregnum of the human intellect:

> In the medieval mind there was no awareness of time, which is ... difficult to grasp. Inhabitants of the twentieth century are instinctively aware of past, present, and future... [But] life then revolved around the passing of the seasons and such cyclical events as religious holidays, harvest time, and local fetes... Generations succeeded one another in a meaningless, timeless blur. In the whole of Europe, which was the world as they knew it, very little happened. Popes, emperors, and kings died and were succeeded by new popes, emperors, and kings; wars were fought, spoils divided; communities suffered, then recovered, from natural disasters. But the impact on the masses was negligible. This lockstep continued for a [thousand years]. Inertia reinforced the immobility. Any innovation was inconceivable; to suggest the possibility of one would have invited suspicion, and because the accused were guilty until they had proved themselves innocent by surviving impossible ordeals – by fire, water, or combat – to be suspect was to be doomed.[6]

The same sense of changelessness pervades the art of the Medieval period. The images are highly stylized. They are figures devoid of individuality, standing separate and aloof in a world of their own, in conventional relations and poses as timeless and unchanging as if they were the Forms of a Platonic heaven. They exist in absolute space and No-time, usually without any definite connection with one another, and always – and this is important – without any explicit reference to what we would define as a *point of view*. For that matter, the very notion of a point of view can only have relevance in a society open to debate and willing to entertain different opinions. This had been the case in the ancient Greek marketplace but it is inimical to a society of totalitarian absolutism where one point of view – and one only – is acceptable. That is the view of the Church, whose view is also that of God and thus absolute and final and uncompromising. Anything opposed to it must be illusion and the work of the devil.

For the same reason perspective – the method of representing three-dimensional objects on a two-dimensional surface – which the Greeks and Romans had used at

[6]Ibid., pp. 22–23.

Byzantine figures at the Basilica of St. John Lateran, Rome. Photograph by William Sheehan, 2008

least in rudimentary form (they called it the method of "slanting images," which – at least according to Pliny the Elder – had been introduced as far back as the sixth century BC by a Greek painter, Cimon of Cleonai) was also rejected as false and deceptive. According to Ross King:

> Plato had condemned perspective as a deceit, and the Neoplatonist philosopher Plotinus (AD 205–270) praised the flattened art of the ancient Egyptians for showing figures in their "true" proportions. This prejudice against the "dishonesty" of perspective was adopted in Christian art, with the result that naturalistic space was renounced throughout the Middle Ages.[7]

Even before the fall of the Roman Empire in the East to the Turks under Mehmed II in 1453, long-lost texts by the classical Latin authors were turning up among the ruins of Rome:

> Manuscripts were disinterred from where they had lain entombed throughout the centuries. The *Annals* of Tacitus, Cicero's *Orator* and *De oratore*, the poems of Tibullus, Propertius, and Catullus (the lone manuscript of whose work was found stoppering a wine barrel), the

[7]Ross King, *Brunelleschi's Dome: how a Renaissance genius reinvented architecture*. New York: Penguin, 2000, p. 34.

> *Satyricon* of Petronius, the poems of Lucretius, a complete copy of Quntilian's *Institutio oratoria* – all of these shards of ancient Rome ... were recovered in the first decades of the fifteenth century.[8]

After the fall of Constantinople, the recovery of the ancient manuscripts accelerated, as Byzantine scholars, fluent in Greek, migrated across Europe, bearing with them precious Greek texts. Mostly they came to Italy, and there the long-lost texts were studied and gradually absorbed, with the inevitable effect of reintroducing diversity into the human mind by offering ideals of human thought and action different from those insisted upon by the Medieval Church. Meanwhile, the invention of the printing press rapidly made these texts more widely available, and broke the Church's long-standing monopoly on literacy (during the Middle Ages even the kings – Charlemagne, for instance – were illiterate). Admittedly, old habits of thought do not change overnight; men had grown used to bending their necks under the yoke of authority and yielding to the tyranny of texts. At first, the same deference was given to the authority of the newly rediscovered ancients as had been given to Aristotle and the Bible, and very few Italians of the fifteenth century would have dared to hold an opinion for which no authority could be found either in antiquity or in the teachings of the Church. Nevertheless, says Bertrand Russell, even this was "a step toward emancipation, since the ancients disagreed with each other, and individual judgment was required to decide which of them to follow."[9] Frances A. Yates makes the same point: "The great forward movements of the Renaissance all derive their vigor, their emotional impulse, from looking backwards."[10] One thinks of the Renaissance as a period of progress rather than of retrospection; but at least in its early stages – in the Tercento (fourteenth century) and the Quattrocento (fifteenth century) – the Renaissance was not so much a birth but a *rebirth* (which is literally what Renaissance means) of ideas lost during the Middle Ages. The human mind, long closed within the dogmas of the quiet past, now opens up the shutters as of an observatory to gaze upon new vistas, new skies.

Even before the scholars had begun to unearth and study these long-lost texts, painters were exploring new ways of seeing through their art. Ultimately, astronomy is a visual science, based on what can be seen, on observation and experiment. The fond hope of the Greeks – at least of those who followed Plato in regarding sensory experience as the realm of illusion and deception – was that somehow through mathematics the mind could grasp the truth more directly and reliably than through the senses. It was an alluring but ultimately impossible dream. It took a long time to realize that Plato's Realm of Ideas, inspired by mathematical objects such as points, lines, triangles and circles, did not exist in a World Apart shining eternally in the Mind of God. Rather, those mathematical objects reflected the human mind's (ultimately the brain's) ability to abstract and generalize concepts

---

[8]King, *Brunelleschi's Dome*, p. 31.

[9]Russell, *History of Western Philosophy*, p. 483.

[10]Frances A. Yates, *Giordano Bruno and the Hermetic Tradition*. Chicago: University of Chicago Press, 1964, p. 1.

from the spatial forms and relations of actual material bodies. Thus they were not objects of intellectual contemplation alone but things of use. The English philosopher Francis Bacon, whose concept of an empirical or inductive science would mark the fruition of several centuries of confused and hazardous growth from uncertain seeds, looked back at the "unkind and ill-starred divorce and separation" of the empirical and rational faculty which had characterized the Medieval Schoolmen. "This kind of degenerate learning," he wrote,

> did chiefly reign amongst the schoolmen; who having sharp and strong wits, and abundance of leisure, and small variety of reading; but their wits being shut up in the cells of monasteries and colleges; and knowing little history, either of nature or time; did out of no great quantity of matter, and infinite agitation of wit, spin out unto us those laborious webs of learning which are extant in their books.[11]

If the laborious webs of learning of Scholasticism were spun in cells and monasteries, the Renaissance's birthplace was above all in the workshops of artists and artisans, men of the sort that Plato and Plutarch had regarded with disdain. Thought in them was conditioned by practice, and ideas were useless unless they could be composed of paint, applied to canvas, or communicated with the hard edge of the chisel to marble.

Among pioneers of seeing, few stand higher than the Florentine artist Giotto di Bondone (1266/1267 or 1276–1337). According to his first biographer, Georgio Vasari, he was a shepherd who was discovered by the Florentine painter Cimabue drawing pictures of his sheep on the rocks and recruited on the spot to learn painting from the master. More likely, he came from a wealthy family and moved to Florence to work as an apprentice in Cimabue's workshop.

Giotto is remembered by astronomers for having painted the return of Halley's Comet in 1304 in the Adoration of the Magi, one of the scenes of his celebrated cycle in the Scrovegni Chapel in Padua. Vasari said of him that "helped by his natural talent and instructed by Cimabue, in a very short space of time Giotto not only captured his master's own style but also began to draw so ably from life that he made a decisive break from the crude traditional Byzantine style and brought to life the great art of painting as we know it today." To see the difference, one need only compare two paintings in the Uffizi in Florence – Cimabue's Madonna and Child Enthroned with Angels and Prophets (c. 1275/1280), which is still Medieval in form and conception, and Giotto's rendition of the same theme (c. 1310). With Giotto comes a decisive shift. He has introduced nothing less than a different way of seeing, which was, in essence, what the Renaissance was all about.[12] Art historian Leonard Shlain points out that "Giotto was the first artist of record to understand intuitively the benefits of painting a scene as if it were viewed from a stationary point of view that was organized about a horizontal and vertical axis.... As a result, the flat picture writing that had been the style for a thousand years suddenly acquired the third dimension of depth... Giotto's 'proto-perspective' places the

[11] Francis Bacon, *The Advancement of Learning* (1605), bk. I.

[12] Gombrich, *Art and Illusion*; figs. 35 and 36 on p. 61.

Cimabue: Madonna and Child Enthroned with Angels and Prophets (c. 1275/1280). The Louvre. Photograph by William Sheehan, 2009

central focus of the viewer outside and in front of the canvas."[13] Within a generation almost every artist who saw his work could appreciate the advantages of painting or drawing so that all lines of sight coming off the painting converged to form an invisible inverted pyramid, the apex of which was the eye.

Among those who took Giotto's "proto-perspective" to the next level was the Florentine architect, Filippo Brunelleschi (1377–1446), who is best remembered for designing and supervising construction of the immense dome of the cathedral of Santa Maria del Fiore in Florence. According to his fifteenth century biographer, the following event is supposed to have occurred in about the year 1425:

---

[13]Leonard Shlain, *Art and Physics: parallel visions in space, time, and light*. New York: William Morrow, 1991, pp. 49–50.

Giotto: Madonna and Child Enthroned with Angels and Saints (c. 1310). Uffizi Gallery. Photograph by William Sheehan, 2008

[On] a small panel about half a *braccio* [one *braccio* = about twenty-eight inches] square … he made a picture of the church of San Giovanni in Florence [the Baptisterium]. He painted the outside of the church and as much as can be seen at one glance. It seems that to draw this picture he went … inside the central door of S. Maria del Fiore. The panel was made with much care and delicacy and so precisely, in the colors of the black and white marble, that there is not a miniaturist who could have done better…. For … the part representing the sky … Filippo placed burnished silver so that the actual air and the sky might be reflected in it, and so the clouds, that one sees reflected in the silver, are moved by the wind when it blows.

The painter of such a picture assumes that it has to be seen from a single point … and that it has to be seen from the right distance… To prevent the spectator from falling into error … Filippo made a hole in the picture at that point in the view … which is directly opposite to the eye of the spectator… This hole was small … on the painted side, and on the back of the panel it opened out into a conical form … like the form of a woman's straw hat.

> Filippo had the beholder put his eye against the reverse side where the hole was large, and while he shaded his eye with his one hand, with the other he was told to hold a flat mirror on the far side in such a way that the painting was reflected in it... When one looked at it thus, the burnished silver already mentioned, the piazza and the fixing of the point of vision made the scene absolutely real. I have had the painting in my hand and have seen it many times in those days, so I can testify to it.[14]

The Baptisterium in Florence, which figured in Brunelleschi's famous demonstration. Photograph by William Sheehan, 2008

The rules of Giotto's proto-perspective and the principles utilized in Brunelleschi's ingenious demonstration were mathematicized and reduced to the exact rules of a new science by Leon Battista Alberti (1404–1472). An architect who designed the churches of Sant'Andrea at Mantua and San Francesco at Rimini as well as the façade of Santa Maria Novella at Florence, he was the first artist to empirically investigate the laws of perspective. Alberti realized that, when a viewer sees something through a window pane, it is possible to get a correct image of it by tracing its outlines provided only one looks with only one eye and does not move one's head.

---

[14] Antonio di Tuccio Manetti, *The Life of Filippo Brunelleschi* (ca. 1480), quoted in *A Documentary History of Art*, vol. 1, Princeton: Princeton University Press, 1981, p. 171 f. For an enlightening discussion, see S. Y. Edgerton, Jr., *The Renaissance Rediscovery of Linear Perspective*, New York: Harper and Row, 1975, and Paul Feyerabend, *Conquest of Abundance: a tale of abstraction versus the richness of being*, Chicago: University of Chicago Press, 1999, p. 94 f.

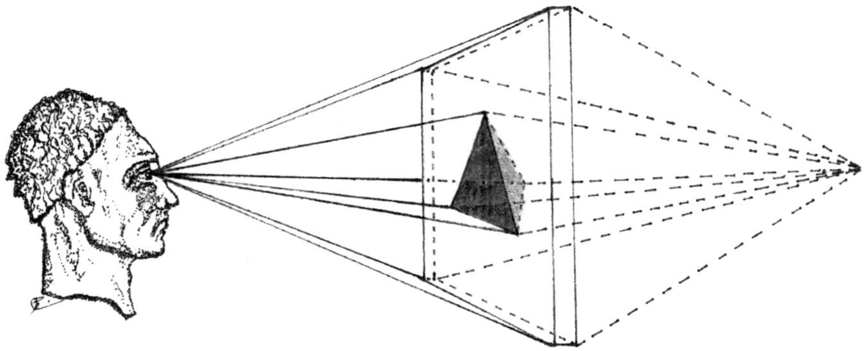

Alberti's diagram illustrating perspective

He then grasped the way to determine the shape and measurements of the visual image of a square of known size at a given distance from the beholder. As we read in his *Vita Anonyma*, he brought about

> things unheard of and that the spectators found unbelievable, and he showed these things [he specialized in paintings of tessellated pavements and checkerboards] through a tiny opening that was made in a little closed box... He called these things "demonstrations," and they were of such a kind that both artists and laymen questioned whether they saw painted things or natural things themselves.[15]

Though it is not known exactly how Alberti first worked out the mathematical laws of perspective, it must have been rather along the lines suggested by William M. Ivins, Jr, in his little book *Art and Geometry*. "As nearly as I have been able to discover," he writes,

> Alberti's procedure – if not in actual manipulation at least in thought – was something like this: He took an oblong box that was long enough for his purpose and removed the top, one end, and one side. In the middle of the remaining end and towards its top he bored an eye or peep hole. On the bottom at the other end he laid a checkerboard just the width of the bottom. When he looked through the eye hole at the checkerboard, he saw that it took the shape of a cross section of a truncated cone. Realizing that the answer to his problem of its apparent height and the apparent comparative lengths of its top and bottom lay in the relationship between the lines of his vision, the eye hole, and the checkerboard, he made a model of the lines of his vision by stretching strings from the eye hole to the corners of the squares on the checkerboard. This enabled him to get out from behind his lines of vision and study them from various positions....[16]

Alberti's book on painting, published in 1435, defined a picture as a "cross-section of the pyramid" formed by the rays extending from the eye to the object. He thus "turned picture-making into a geometrical problem and the painter into a producer

---

[15]Leon Battista Alberti; quoted in William M. Ivins, Jr., *Art and Geometry*, p. 70.

[16]Ivins, *Art and Geometry*, pp. 70–72.

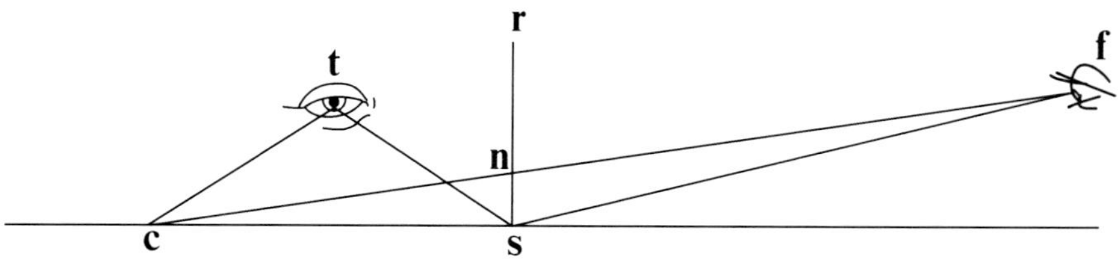

Alberti's construction as rendered by Leonardo

of cross-sections of optical pyramids."[17] It had a profound influence on Leonardo da Vinci (1452–1519), who also "threatened the certitude that knowledge had been forever fixed by God, the rigid mind-set which left no role for curiosity or innovation."[18] For Leonardo, too, art was chiefly about *saper vedere* – knowing how to see. He began the manuscript of his book on painting with a kind of manifesto in which he asserts his resolve to rely not on the dictates of authority but on his own experience:

> I am fully conscious that, not being a literary man, certain presumptuous persons will think that they may reasonably blame me; alleging that I am not a man of letters.... They will say that I, having no literary skill, cannot properly express that of which I desire to treat; but they do not know that my subjects are to be dealt with by experience rather than by words....[19]

Characteristically, Leonardo begins with first principles, and takes as his starting-point the eye itself:

> Behold here O reader! A thing concerning which we cannot trust our forefathers, the ancients, who tried to define what the Soul and Life are – which are beyond proof, whereas those things, which can at any time be clearly known and proved by experience, remained for many ages unknown or falsely understood. The eye, whose function we so certainly know by experience, has, down to my own time, been defined by an infinite number of authors as one thing; but I find, by experience, that it is quite another.[20]

"Look for yourself," he says. Belief must be based henceforth not on the blind faith which had "permitted a mafia of profane popes to desecrate Christianity."[21] The truth is not the province of those "who do not see, and yet believe." Rather, they are even more blessed who look and believe only in what they can see and demonstrate for themselves.

That, in a nutshell, is the difference in the sensibility of the Renaissance and the Middle Ages. It would change forever the way that man looked at the universe – though

[17] Feyerabend, *Conquest of Abundance, p. 96.*

[18] Manchester, *A World Lit Only by Fire*, p. 91.

[19] Leonardo da Vinci, The Notebooks of Leonardo da Vinci; edited by J. P. Richter. New York: Dover, 1970, reprint of 1883 edition, p. 14.

[20] Ibid., p. 19.

[21] Manchester, *A World Lit Only by Fire*, p. 91.

the texts would remain powerful still, and even today the triumph of experience over authority is far from assured.

An inheritor of the legacies of Giotto and Brunelleschi, Alberti and Leonardo, Nicolas Copernicus – born at Thorn, on the River Vistula, in 1473 – became the first man to grasp that the laws of perspective could be applied not only to frescoed walls and ceilings but to the vault of cosmos itself.

Nicolas Copernicus. Portrait in the Town Hall, Thorn. Photograph by Tomasz Mazur

Copernicus's father, a prosperous wholesale merchant, died when Nicolas was only ten, and he passed into the care of his uncle, Lucas Watzelrode, the bishop of the ecclesiastical state of Ermland (a See in the north of what is now Poland). During Copernicus's lifetime, nepotism was rampant in the Church. Copernicus himself benefited from his uncle's powerful position when, as a 24 year old mathematics student at the University of Cracow, he was appointed a canon (not a priest; he was never ordained) in the Cathedral at Frombork (or Frauenburg), a small town 40 km east of Gdansk and overlooking the Frisches Haff, a lagoon of the Baltic. His future was thus secure. He was not, however, eager to settle into the position, and began a series of postponements and evasions of his duties while he pursued further studies in Italy, still in the flush of its glorious Renaissance.

He remained there for 10 years, years which must have been immensely stimulating for someone as intelligent and wide-awake as Copernicus clearly was. It was the best of times and the worst of times. The papacy was then at the absolute nadir of its moral authority under the worldly and corrupt Renaissance popes, of whom arguably the most notorious was Alexander VI Borgia, who occupied the Throne of St. Peter from 1492 to 1503. Elected by bribery, Alexander's many mistresses included Vannoza dei Cattanei, who bore him Cesare and Lucrezia Borgia. The face of another mistress was bestowed on the Blessed Virgin in a scene painted in Alexander's bedroom, only recently recovered after having been whitewashed by one of Alexander's more pious successors in the eighteenth century. Though his sexual orgies were notorious, even worse were his schemes to promote his family's interests (a subject which forms one of the chief topics of Niccoló Machiavelli's *Prince*). Ultimately, he centered his hopes on the ruthless figure of his son, Cesare, who by policies of siege and assassination was able to bring northern Italy under his control and was on the verge of success when he suddenly and unexpectedly died. To pay for these adventures, Alexander resorted to the sale of indulgences, a practice which greatly inflamed resentment among the restive peasants of Germany. Yet Alexander was not entirely evil; he was also a patron of the arts, as were his no less saintly successors, the soldier-pope Julius II (pope from 1503 to 1513), for whom Michelangelo painted the Sistine Chapel, and Leo X (1513–1521), a member of the powerful Florentine Medici family, who conferred great commissions on Michelangelo's rival Raphael. All of these grandiose artistic and building projects (which included the new St. Peter's) had to be paid for somehow. So the indulgences continued unabated, and the German peasants groaned. In the end it was Leo X who would excommunicate a defiant German monk, Martin Luther, whose 95 theses, nailed to a church door in Wittenberg in 1517, threw down the gauntlet of the Protestant Reformation.

As these events swirled, Copernicus studied. In 1496 we find him in Bologna, at Europe's oldest university, where he became a disciple of Domenico Maria de Novara. From Novara he absorbed a sense of just how unaesthetic and objectionable was the Ptolemaic system, with its "pedantic apparatus of epicycles, deferents, and equants." He made his first astronomical observation, an occultation of the star Aldebaran by the Moon, on March 9, 1497.

During the Jubilee Year 1500, he was in the Rome of Leonardo and Michelangelo, lecturing on mathematics. He attended the University of Padua – whose faculty would later include Galileo – for the study of medicine; the University of Ferrara where he went for his doctorate in canon law; then back to Padua to resume his medical studies. Finally, in 1506, he returned to Poland, and spent several years as his uncle's personal physician at the Episcopal palace at Heilsburg. Not until his uncle's death in 1512 did he take up his long-deferred residence at Frombork.

By then he was almost 40.

It is sometimes claimed that temperamentally Copernicus was a conservative. Certainly his long and desultory preparation shows him to have been more deliberative than impetuous. Arthur Koestler famously referred to him as the "timid canon"[22] in attempting to account for the long delay in the development of his ideas and the fact that he waited until the very end of his life to publish. But despite his living in a Tower, it was not an Ivory Tower.

As canon, Copernicus was entrusted with many harassing duties, and his burdens were heavier than is sometimes assumed. He kept all the accounts of the diocese,

[22] Arthur Koestler, *The Sleepwalkers*. New York: Grosset & Dunlap, 1959.

Occultation of Delta Geminorum by Mars, April 7, 1976. Painting by Julian Baum based on observations by William Sheehan with a 6-in. reflector and an eyepiece magnifying 250×. © Julian Baum

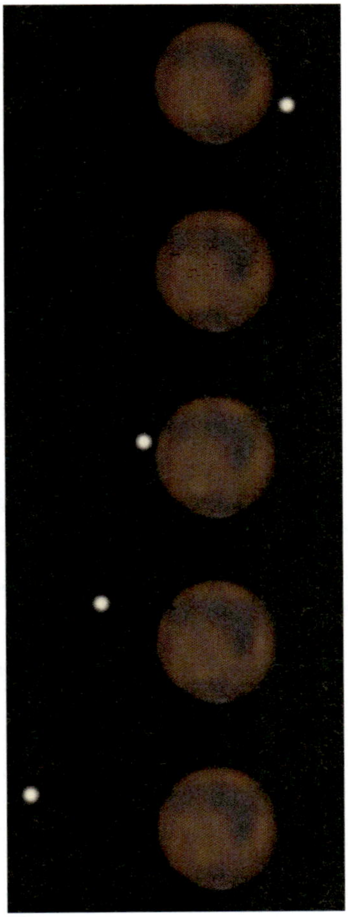

Frombork Cathdral. Photograph by Tomasz Mazur

Copernicus's tower. Photograph by Tomasz Mazur

saw that the chapter's political interests were protected, collected rents and settled disputes for the peasants who worked the fields surrounding the Cathedral compound. His medical skills were in constant demand. He waged war against the Teutonic knights, and helped plan the reconstruction of Ermland after the war ended in 1521. The next year he was called on by Pope Hadrian VI to make recommendations for the reform of the calendar, and presented a plan for the reform of the currency to the Diet of Graudens. One has the impression of a man of great physical and moral courage and inexhaustible energy. A few years ago, when his remains were unearthed from an unmarked grave in the Cathedral of Frombork, he was found to have had a broken nose. One pictures him as someone who, while preferring diplomacy to violence, could stand his ground whenever the situation required it.

Nevertheless, he did find time for his astronomy, and it was clearly no mere hobbyhorse for him. In existing portraits, all of which seem to have been based on a self-portrait (he dabbled in art along with anything else), he wears the impassive mask of a man guarded and controlled. But the mask conceals a burning passion, as well as a secret plan to devote himself to the grandiose project of reforming astronomy from the moment he returned from Italy. He did not set out to be a brash revolutionary; he was never an iconoclast as Giordano Bruno would be. Respectful of authority, he remained a faithful son of the Roman Catholic Church. To have been otherwise in that age, for a man in his position, was inconceivable. But he was also insatiably curious, and he devoured every text he could lay his hands on.

Scholasticism during the thirteenth century had done much to enlighten mankind by introducing the works of Aristotle into the West. But, as J.L.E. Dreyer rightly observes, "no amount of study of Aristotle or of the scholastic writers could itself advance science."[23] The Medievals had tended to see all things from a common peephole, and with a shared dogmatic unity based on St. Thomas Aquinas's architectonic reconciliation of Aristotle and the Bible. In the *Almagest*, Ptolemy had expressed the impossibility – even ridiculousness – of the Earth's motion. He had remained satisfied with what was essentially a God's-eye view in which the terrestrial platform, absolutely at rest in the center of the universe, was the only vantage point conceivable. The Earth was not only the privileged reference frame for regarding the movements of the planets, it remained the perfect metaphor for Medieval Christendom in which everything was at rest and in its proper place in a universe as well-balanced and carefully supported as a Medieval Cathedral. Henry Adams writes:

> Truth, indeed, may not exist; science avers it to be only a relation; but what men [once] took for truth stares one everywhere in the eye and begs for sympathy. The architects of the twelfth and thirteenth centuries took the Church and the Universe for truths, and tried to express them in structures which should be final. Knowing ... precisely where the strains were to come, they enlarged their scale to the utmost point of material endurance, lightening the load and distributing the burden until the gutters and gargoyles that seem mere ornament, and the grotesques that seem rude absurdities, all do work either for the arch or for the eye; and every inch of material, up and down, from crypt to vault, from man to God, from the Universe to the Atom, had its task, giving support where support was needed, or weight where concentration was felt, but always with the condition of showing conspicuously to the eye the great lines which led to unity and the curves which controlled divergence; so that, from the cross on the flèche and the key-stone of the vault, down through the ribbed nervures, the columns, the windows, to the foundation of the flying buttresses far beyond the walls, one idea controlled every line.[24]

Of course, remove the seemingly solid platform – unsettle the Earth – and the whole Cathedral of Medieval thought must shake and teeter and fall. The idea that began that shaking is the one Adams refers to – truth as a relation; it begins with the Renaissance artists and advances with Copernicus. As soon one tries to see how things appear from a multiplicity of viewpoints – see, as it were, through different peepholes – they can never again appear as settled as they had seemed. It is now permissible to vary the position of the observer and to ask: How do things look from *there* instead of from *here*? In particular, how do things look when one looks at the planets not from the surface of the Earth but from some point in the midst of all of them? Possessed of what might be described as planetary *empathy*, the ability to "see ourselves as ithers see us," the relations are changed and – the most astounding

[23] J. L. E. Dreyer, *A History of Astronomy From Thales to Kepler*, p. 281.

[24] Henry Adams, *Mont St.-Michel and Chartres*, p. 358.

consequence of all – at one stroke the apparently flat convex surface of the heavenly vault that so long seemed to hang like a fiery canopy over the Earth opens up along a third dimension into a dizzying and potentially limitless vastness.

Already as Novara's student in Bologna, Copernicus had heard expressions of dissatisfaction about the pedantic apparatus of the Ptolemaic system. When he began to inquire whether other views had existed among the ancients, he found:

> Our ancestors assumed, I observe, a large number of celestial spheres ... to explain the apparent motion of the planets.... The planetary theories of Ptolemy and most other astronomers, although consistent with the numerical data, seemed ... to present no small difficulty. For these theories were not adequate unless certain equants were also conceived; it then appeared that a planet moved with uniform velocity neither on its deferent nor about the center of its epicycle. Hence a system of this sort seemed neither sufficiently absolute nor sufficiently pleasing to the mind.
>
> Having become aware of these defects, I often considered whether there could perhaps be found a more reasonable arrangement of circles, from which every apparent inequality would be derived and in which everything would move uniformly about its proper center, as the rule of absolute motion requires.[25]

What, then, should be the proper center for such a rational system of circles? Might it be the Sun?

Copernicus must have been reassured to find precedents for his idea among the ancients. As he discovered, there were several references in ancient texts to heliocentric ideas, of which the most notable, of course, refer to Aristarchos. For a mind trained to respect authority – a careful, lawyerly, canonical mind – this was important. Authority must be opposed by other authority. It was also characteristic of a Renaissance man to look backward before leaping forward. But Copernicus also seems to have had other reasons – emotional and aesthetic – for choosing the Sun rather than the Earth as his "watchtower." As he later explained in a remarkable passage of *de Revolutionibus*:

> In the middle of all sits the Sun enthroned. In this most beautiful temple, could we place this luminary in any better position from which he can illuminate the whole at once? He is rightly called the Lamp, the Mind, the Ruler of the Universe: Hermes Trismegistus names him the Visible God, Sophocles' Electra the All-Seeing. So the Sun sits as upon a royal throne, ruling his children, the planets which circle round him.[26]

---

[25] Nicolas Copernicus, *Commentariolus*, in Edward Rosen, *Three Copernican Treatises* (New York: Dover, 1959 reprint of 1939 edition), p. 57.

[26] Copernicus, *Revolutions of the Heavenly Spheres*, Book I, chapter 10. Alexandre Koyré points out that "it is not always, or perhaps not sufficiently, appreciated that by placing the Sun at the centre of the Universe in virtue of its dignity, Copernicus returned to the Pythagorean conception and completely overthrew the hierarchy of positions in the ancient and medieval Cosmos, in which the central position was not the most honourable, but, on the contrary, the most unworthy. It was in effect, the *lowest*, and consequently appropriate to the Earth's imperfection. Perfection was located above in the celestial vault, above which were 'the heavens' (Paradise), whilst Hell was deservedly placed beneath the surface of the Earth." *The Astronomical Revolution: Copernicus–Kepler–Borelli*, translated by R. E. W. Maddison. New York: Dover, 1992 reprint of 1961 edition, p. 115.

Aristarchos, as we know, had somehow managed to grasp the fundamental idea of the Earth moving around a stationary Sun as early as the third century BC. Unfortunately, we can only speculate as to what his reasoning was, but it most assuredly was a *tour de force* – rather like the remarkable treatise on determining the distances of the Sun and Moon (the only one of Aristarchos's works to survive) or Archimedes' heroic computation of the volumes of cylinders and spheres. The Medieval mind had long been used to thinking of the Earth as situated underneath – rather than in – the heavens, and for over a thousand years, the planets, whose movements were believed to be due to the impetus or motor force given them by angels, looked down from above no less inscrutably than the iconic figures of the Byzantine mosaics:

> Images such as these, looking down on us from the golden, glimmering walls, seemed to be such perfect symbols of the Holy Truth that there appeared to be no need ever to depart from them.[27]

Seen instead from the view of Giotto's proto-perspective or Alberti's laws of perspective, however, it became possible to rethink astronomy by taking up a position *in* rather than *under* the heavens, and to vary at will the position of the observer.

Sometime probably while he was still in Italy, Copernicus had performed this feat mentally. Perhaps he did even more. He was, after all, trained in art; I can imagine him trying the experiment of Alberti with his checkerboard, taking an oblong box with its eye hole and stretched strings and getting out from behind his lines of vision in order to study them from various positions. *Here is the view from the Earth, there is the view from the Sun.* Instead of remaining complacently in place at the peephole in back of the box, Copernicus could move about and regard the planetary system from any position he wished.

And what would things look like if viewed from the peephole placed in the All-seeing one, the Sun? That was the critical experiment. As soon as Copernicus moved the center to the Sun, the strings between the Earth and the other planets magically disentangled themselves and revealed that the so-called second inequality in the motion of each planet – its apparent backward arc or retrograde motion – for which Ptolemy had been forced to introduce the largest epicycle, was a mere perspective-illusion of the Earth's own movement around the Sun. The diagram that appears here sorted itself out of those strings. Thus, at a stroke, Copernicus was able to eliminate the main epicycle of each planet. According to Leonard Shlain:

> In his flash of insight, belief in the previous system was doomed. The hub of the solar system was the Sun, he realized. Copernicus, stepping outside the existing model of the solar system and looking back on it from an imaginary outside perspectivist point of view, was able to rearrange the planets and the sun in an entirely new way. His revolution achieved for the space of science what Giotto's perspective had done for the territory of art. The "underdimensional" medieval worldview was expanded to encompass a larger richer third dimension of depth.[28]

[27] E. H. Gombrich, *The Story of Art*, New York: E. P. Dutton, 13th ed., 1978, p. 101.

[28] Leonard Shlain, *Art and Physics*, p. 59.

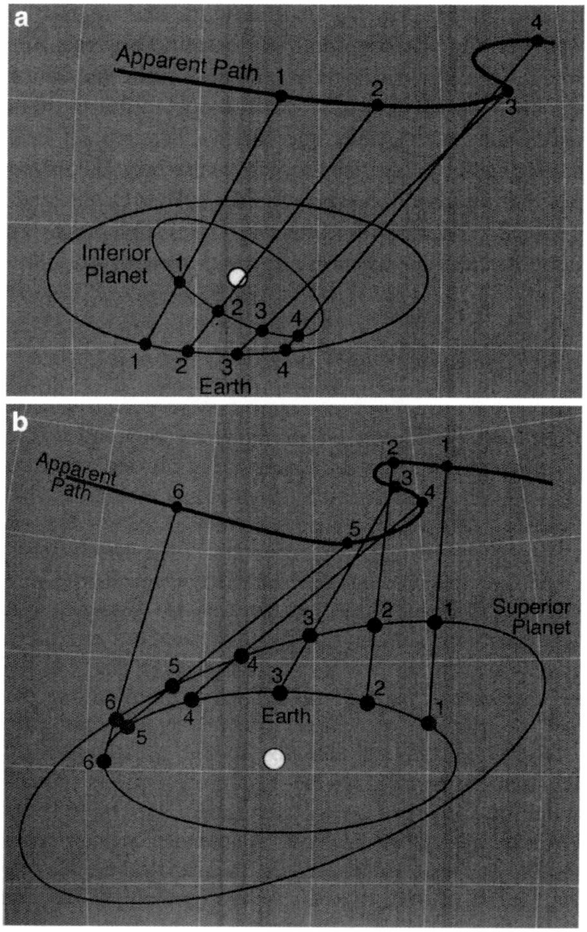

Retrograde movements according to Copernicus

Effectively this was true. However, a mere flash of insight was hardly enough to free the human mind from the oppression of a thousand years cramping orthodoxy – anymore than the religious questions which preoccupied men of his time were solved once and for all by the stand Luther took on behalf of the individual conscience a few years later at Wittenberg. (Ironically, Luther himself would mock Copernicus as "the fool who wanted to turn the whole art of astronomy on its head," and dismiss him by asserting: "As Holy Scripture tells us, Joshua bid the sun to stand still and not the earth.")

Sudden flashes of insights of the kind I have described involve the visual Right Hemisphere, and are usually inarticulate. Copernicus now had to put what he had glimpsed into mathematical demonstrations that would break through the heavy weight of trappings and the sheer inertia of what had become ingrained and settled "truth." Daniel Boorstin points out:

> Astronomers [had become] adept at explaining away what seemed only minor problems by a variety of complicated epicycles, deferents, equants, and eccentrics, which gave them a heavy vested interest in the whole scheme. The more copious this peripheral literature became, the more difficult it became to retreat to fundamentals. If the central scheme was not correct, surely so many learned men would not have bothered to offer their many subtle corrections.[29]

What Copernicus undertook to do was, Boorstin adds, nothing less than attempt to "displace a system that was amply supported by everyday experience, by tradition, and by authority." It must have seemed daunting. He was about to give a new meaning to the word revolution. But so far it was a silent revolution.

Copernicus had begun to prepare himself for the work almost immediately on arriving in Frombork. We know that he planned to make observations, for in the Cathedral Records for April 1513 appears the note: "Doctor Nicolaus paid into the treasury of the chapter for 800 bricks and a barrel of chlorinated lime from the Cathedral work-yard."[30] With these he built – or more likely commissioned someone to build for him – a viewing platform. On a visit Dava Sobel was shown the place just outside the Cathedral wall where Copernicus is thought to have had his house (the quarters in the Northwest Tower would have served as his office but his residence only during war); she also saw the place, in the field between his house and the wall, where the brick platform stood. On this brick platform he would have deployed a large triquetrum – a parallactic ruler, used since Ptolemy's day to measure the altitudes of celestial bodies – and other instruments. Though he made some observations of his own, he was hardly an adept; he would later remark that he would have been as happy as Pythagoras was on discovering his theorem to have observations accurate to within 10 min of arc. That degree of accuracy, though not perhaps of happiness, was only achieved later – by Tycho.

Perhaps even as the brick platform was rising next to the Cathedral wall, Copernicus was writing his short sketch – *Commentoriolus*, or the little commentary – of a new planetary theory based on the proposition that the Earth, instead of being the center of the universe, is an ordinary planet, in orbit around the Sun. (It was never published during his lifetime, and only a few copies of it circulated among his friends in manuscript; one is noted in the catalog of one Mathias de Miechow of Cracow in 1514, from which we know that it existed by then). Already Copernicus realized that by adopting the heliocentric artifice he could achieve a remarkable simplification of the planetary system. To put it in perspective: when Giovanni de Dondi, a Paduan clockmaker who based his designs on Ptolemy, constructed a famous clockwork of the heavens in 1350, he had no choice but to introduce a separate mechanism for each of his planets. Thus, his clock had seven faces or dials, each carrying a planet. The work cost him 16 years of tinkering. In its time, it was a work of technical sophistication as great as John Harrison's marine chronometer would be in the eighteenth century, and succeeded in tracing the movements of the planets with an accuracy equal to the observations then being made. Dondi's clock reflected the construction

---

[29] Daniel Boorstin, *The Discoverers*, pp. 295–296.

[30] Dava Sobel, personal communication; March 3, 2009.

of the planetary system as understood in his time. By contrast, Copernicus's scheme – by eliminating the large epicycle for each planet – could represent the motions to the same level of accuracy, but by means of a single clockface.

In the most remarkable passage in *Commentariolus*, Copernicus explains with perfect lucidity the way the heliocentric theory disposes of the retrograde motions of the planets, especially Mars, which had so long perplexed astronomers:

> There is a second inequality, on account of which the planet seems from time to time to retrograde, and often to become stationary. This happens by reason of the motion, not of the planet, but of the earth changing its position in the great circle. For since the earth moves more rapidly than the planet, the line of sight directed toward the firmament regresses, and the earth more than neutralizes the motion of the planet. This regression is most notable when the earth is nearest to the planet, that is, when it comes between the sun and the planet at the evening rising of the planet....[31]

Copernicus's model threw into highlight a number of features of the Ptolemaic theory that previous astronomers had simply glossed over as accidental and adventitious. In Ptolemy's scheme, the centers of the epicycles of the outer planets move along a large circle, known as the deferent, in a period essentially that of the planet about the Sun measured with respect to the fixed stars: for Mars, 687 days, for Jupiter, 11.9 years, for Saturn, 29.5 years. At the same time the planet moves about its epicycle, in the opposite sense, in the time of its synodic period. The mean values are: for Mars 780 days, for Jupiter 399 days, for Saturn 378 days. When the planet is on the near side of its epicycle to the Earth, its motion in the epicycle takes place in a direction contrary to that with which the epicycle is carried along the deferent, so that the planet appears to be moving retrograde. Since these retrogressions occur near opposition, the motion in the epicycle has to be synchronized so that the line from the center through the Earth intersects the Sun. For the inner planets, the motion about the Earth of the center of the epicycle is 1 year for both, and the line from the Earth through the center of the epicycle always passes through the Sun.

Now in a geocentric system, there is no obvious reason why the Sun should be so intimately involved with all of these constructions. There, these pointings and alignments must be regarded as mere coincidences. By considering the way things looked not from the Earth but from the Sun, however, Copernicus could rationalize these features: they now appeared as reflections of the Earth's motion among the other planets and around a common center, the Sun.

Copernicus regarded *Commentariolus* as only a sketch. What he now needed to do was produce what would be, in effect, a new *Almagest*. He had to carry out the massive mathematical complications that would generate a theory that would satisfy the observations as well as Ptolemy's.

Though Copernicus put the Earth in rotation on its axis and in motion around the Sun, for the rest he still adopted the Ptolemaic machinery. The main epicycle of Ptolemy for each planet was eliminated but lesser epicycles had to be introduced to account for small irregularities ("inequalities") that had been accumulating

[31]Copernicus, *Commentariolus*, p. 77.

since before the time of Apollonius. The Aristotelian crystalline spheres (*orbes*) also survived, appearing prominently in the title of Copernicus's great book, *de Revolutionibus orbium caelestium*, where the orbs are not the planets themselves but the revolving spheres on which they are presumed to move. All the while he was exquisitely sensitive to the fact that in presenting his theory of the moving Earth he was arguing for a position that – at least on the face of it – would seem ridiculous. He had finished the manuscript by the early 1530s, but hesitated to bring it out. He was afraid, he says, that others would "shout to have me and my opinion hooted off the stage."

In the end, it was only the entreaties of his friends – and especially the persuasions of an enthusiastic young student, Georg Joachim, who called himself Rhaeticus, that prevailed. Rhaeticus was a Lutheran, no less, who in 1539 came to Frombork from Wittenberg to learn from Copernicus about his theory. Rhaeticus soon produced a First Report (*Narratio Prima*) of Copernicus's theory, and printed it at (now Gdansk, Poland) in 1540. It quickly sold out, and a second edition was printed within a year.

Paul III and his nephews (Ottavio Farnese, Duke of Parma, and Alessandro, in Cardinal's garb), by Titian. The Gallerie di Capodimonte, Naples

With this encouragement – and the solid support of several influential cardinals – Copernicus at last turned to preparing his own work for the press. Preceded by a solemn and dignified preface, he dedicated the book to Pope Paul III.

The Roman Catholic Counter-reformation against Protestantism was by now underway. Paul himself had revived the dread Roman Inquisition as well as convened the Council of Trent, which eliminated the most egregious abuses in the Church that had occurred under the Renaissance popes. At the moment – despite the laying down of more entrenched lines in religion between Catholics and Protestants – Christian dogma still did not mandate opinions as to the way things were arranged in the heavens, and Copernicus himself was never personally in danger. By the time the book went to press – it was printed by a Lutheran theologian, Andreas Osiander, in Germany – Copernicus was mortally ill; the first copies arrived when he was literally on his deathbed. If he was even conscious when he received them, he would have been incensed to discover that a spurious preface – unsigned, and actually written by Osiander himself though presented in such a way as to suggest it represented Copernicus's own views – had been added as front matter to the book. Without authorization, Osiander had boldly asserted:

> Since the fame of the new hypotheses of this work (according to which the earth moves and the sun stands still and fixed at the center of the universe) is already widespread, I do not doubt that some learned men feel themselves very much offended, deeming this a starting-point for throwing into confusion the liberal arts, until now so well and so long set in order. But if they care to consider the matter more closely, they will find that the author is not worthy of blame, since it is proper for astronomers to formulate diligently and skillfully the true order of celestial motions. Since for various reasons they cannot find the true causes of the latter, it is permissible to feign and fashion them as they please by means of geometrical principles, through which it is possible to calculate past and future motions. For this reason, these hypotheses do not need to be true or even probable.... [So] let us assay these hypotheses only for their wonderful and inventive ease of calculation, for if someone takes them as true, he will leave this discipline more foolish than if he had never entered it.[32]

Osiander's position is more simply restated by Edward Rosen: "Since divine revelation is the only source of truth, astronomical hypotheses are not concerned therewith, and serve only as a basis of calculations."[33] This was to be the position of the Churchmen who would oppose Galileo. Unaware of the imposture (it was unmasked by Giordano Bruno who, without fingering Osiander, recognized it as the work of an "ignorant and conceited ass," and finally proved by Kepler), they would ascribe it to Copernicus himself. However, no one who had actually read *de Revolutionibus* could possibly attribute such position to him. Clearly, Copernicus had asserted the

[32]Quoted in Rosen, *Three Treatises*; pp. 24–25. Osiander is rehashing the argument of Simplicius, the sixth century AD author of a commentary on Aristotle's "On the Heavens." Simplicus had argued that the astronomers had not demonstrated their hypotheses, since the same phenomena are sometimes explained by different hypotheses.

[33]Ibid., p. 24.

doctrine of the Earth's motion as the most important innovation of his system, and regarded it as a physical reality.

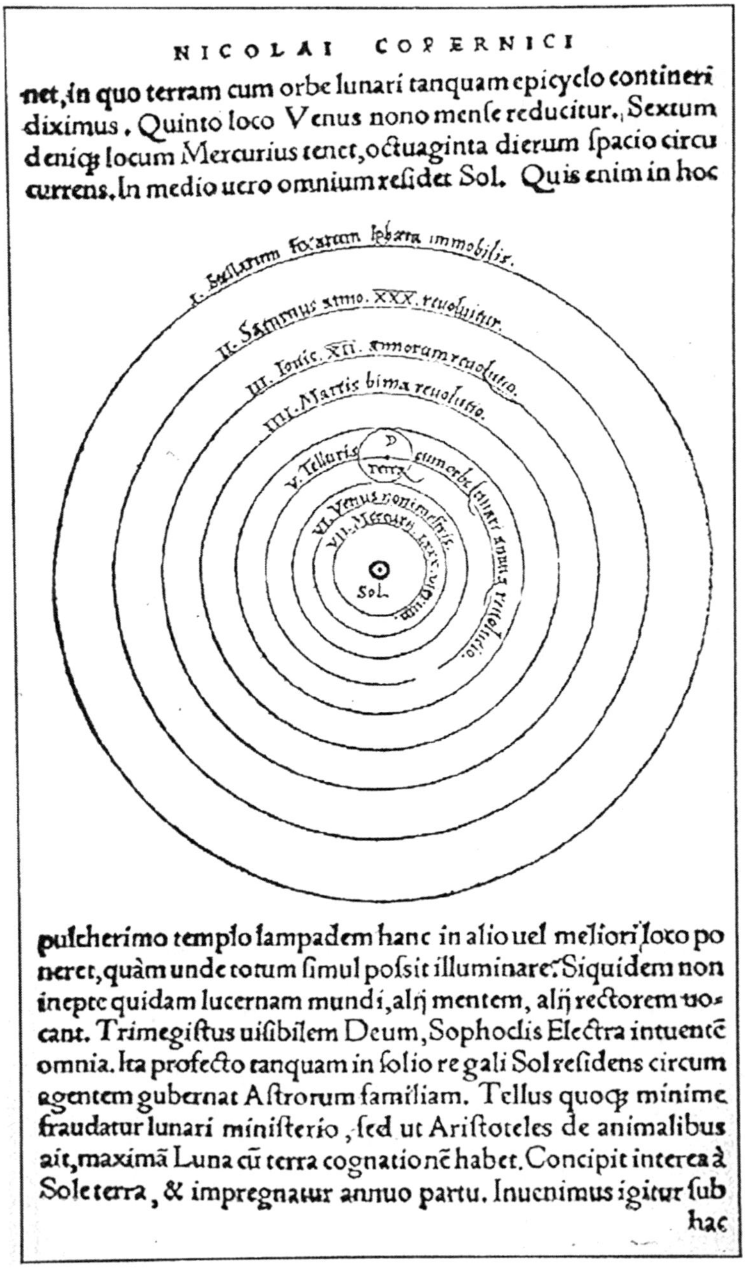

NICOLAI COPERNICI

net, in quo terram cum orbe lunari tanquam epicyclo contineri diximus. Quinto loco Venus nono mense reducitur. Sextum deniq; locum Mercurius tenet, octuaginta dierum spacio circu currens. In medio uero omnium residet Sol. Quis enim in hoc pulcherimo templo lampadem hanc in alio uel meliori loco poneret, quàm unde totum simul possit illuminare? Siquidem non inepte quidam lucernam mundi, alij mentem, alij rectorem uocant. Trimegistus uisibilem Deum, Sophoclis Electra intuentē omnia. Ita profecto tanquam in solio regali Sol residens circum agentem gubernat Astrorum familiam. Tellus quoq; minime fraudatur lunari ministerio, sed ut Aristoteles de animalibus ait, maximā Luna cū terra cognationē habet. Concipit interea à Sole terra, & impregnatur annuo partu. Inuenimus igitur sub hac

Copernican system as rendered in Copernicus's great book, *De Revolutionibus* (1543)

By applying the method of scientific perspective to the motions of the planets, Copernicus had opened up the deeps of the heavens. The distance to the "fixed stars" was now immense; instead of consisting of a series of internested spheres without any waste space between, Copernicus implied the existence of a vast space between the sphere of Saturn and that of the Fixed Stars – an implication that was first fully grasped by the peregrinating Dominican friar-turned-heretic, Giordano Bruno.

The son of a professional soldier, Bruno was born at Nolan, near Naples, about 1548, and often referred to himself simply as "the Nolan." He entered the Neapolitan convent of San Domenico at the age of 14 and soon distinguished himself for his prodigious feats of memory. He was ordained a priest and licensed to teach theology, but almost at once was under investigation by the local head of the Dominicans for expressing unorthodox views (he once scoffed at a fellow novice for reading a devotional poem about the Blessed Virgin). By 1576, he had grown so uncomfortable in Naples that he decided to escape – first to Genoa, where he abandoned his clerical garb and taught Latin and astronomy. Thus he embarked on a restless career in which he would wander across vast regions of the art of memory, hermetic philosophy, and magic, and travel across Europe trying to find a hospitable reception to his unconventional ideas.

From Genoa, he went to Venice; from Venice to Padua; from Padua to Lyons; from Lyons to Calvinist Geneva, where he adopted Calvinism but was jailed for attacking a local philosopher; from Geneva to Toulouse, where he lectured on Aristotle and astronomy; from Toulouse to Paris, where he taught Henry III about the art of memory and wrote a play which cannot have been very successful, since on the title page he referred to himself as "Bruno the Nolan, the Academic of no Academy; nicknamed the exasperated." In the company of the French ambassador, he visited England in 1583, and hobnobbed with Sir Philip Sidney, the poet, and other Elizabethan worthies in London. Through them, some of his ideas seem eventually to have filtered to Shakespeare. Having early embraced Copernicanism, he lectured at Oxford in favor of the heliocentric theory against the Oxford Ptolemaicists. His adversaries distinguished themselves mainly by their pedantry but succeeded in discrediting him by showing that he did not even have the details of the Copernican system right. Nevertheless, despite being muddled in the details, his vision was sublime: he derived an infinite universe from the heliocentric theory. Not only was the Earth no longer at the center of the universe, claimed Bruno; neither was the Sun – it was only one of infinitely many centers in a universe of limitless dimensions. "There is in the universe neither center nor circumference," he wrote.

Clearly he was blazing a dangerous path into realms of ideas that were bound to get him in trouble with the Church. If the Medieval universe had represented an architectonic whole of which, in Henry Adams's words, its architects had known "precisely where the strains were to come, ... enlarged their scale to the utmost of material endurance," this infinite universe of Bruno's was obviously more than any combination of flying buttresses or other props and supports could possibly bear. The vast space beyond Saturn could not possibly be stuffed within the closed walls of the Medieval universe.

In 1591, Bruno fatefully decided to return to Italy. He was befriended by a Venetian nobleman, Giovanni Mocenigo, and took up residence in the Palazzo Mocenigo on the Grand Canal. Mocenigo hoped that Bruno would improve his memory but apparently was disappointed; moreover, he suspected that a relationship was developing between Bruno and his wife. They fell out, Mocenigo took revenge by betraying Bruno to the Church authorities, and Bruno spent the last 6 years of his life in the miserable dungeons of the Roman Inquisition. In the end, he was condemned under Clement VIII for eight heretical propositions (we shall never know for sure what they were; the documents that contained the statement of these propositions were destroyed). With a spike driven through his tongue to silence it, he was burned alive as a heretic in the Campo de'Fiori in Rome on February 17, 1600, exactly 364 years to the day before I merrily went splashing through puddles on the way home from school and received – it might well have been from the spirit of Bruno himself – that flash of insight that the Sun was a star and that all the stars were suns, each with other planets, and so worlds without end. Amen.

Chapter 7
A Passion in Bohemia

O curas hominum, o quantum est in rebus inane.
(O the cares of man, how much of everything is futile.)

Aulus Persius Flaccus (AD 34–62); adopted by Johannes Kepler as his motto

Shakespeare's *Hamlet* has always merited a footnote in astronomical histories. The setting of the play is Elsinore, the site of the fifteenth century castle greatly enlarged during the 1580s on the Öresund (Sound) between Denmark and Sweden. It is on the other end of the Baltic from Copernicus's Frombork and not far from the Baltic island of Hven, where Tycho Brahe – a legendary Dane in his own right – built the most splendid observatory of the age. Though Hamlet speaks a few lines such as, "O God, I could be bounded in a nutshell and count myself a king of infinite space," which could have been inspired by Bruno, Tycho is even more tangibly alluded to in the play. The names of Hamlet's two school friends, Rosencrantz and Guildenstern, bear the names of Tycho's ancestors as recorded on the famous portrait Tycho affixed as the frontispiece of his *Astronomical Letters* of 1596. I can never see Hamlet without thinking of Tycho.

Tycho's father, Otto, was governor of Helsingborg Castle on the other side of the Sound from Elsinore. Before Tycho's birth, at Knudstrup, Denmark, in 1546, he entered into a strange agreement with his brother Joergen; if he had a son he would give him up so the latter could adopt him and raise him as he own. Though Otto began to regret the pact, Joergen was not to be denied – he abducted the child and took him to his own castle at Tostrup. After the initial shock died down, Joergen was allowed to keep him.

Tycho was none the worse off for it. Joergen was extremely well-to-do; he doted on Tycho, and could afford to give him the best education money could buy. When Tycho was 13, he was sent to the University of Copenhagen to begin the study of the law, but soon afterwards an event occurred that would forever change the direction of his life. On August 21, 1560, an eclipse of the Sun took place just as astronomers had predicted. According to Tycho's early biographer Pierre Gassendi, "Tycho thought of it as something divine that men could know the

W. Sheehan, *A Passion for the Planets: Envisioning Other Worlds, From the Pleistocene to the Age of the Telescope*, DOI 10.1007/978-1-4419-5971-3_7,

Tycho Brahe's portrait in *Astronomiae instauratae mechanica* (Wandesburg, 1598). This handcolored engraving may well be the best likeness of him, showing that the disfigurement sustained in his youthful duel probably included severance of a branch of the facial nerve

motions of the stars so accurately that they could long before foretell their places and relative positions."[1]

His interest aroused, Tycho purchased a copy of Ptolemy's *Almagest*, and worked his way through it. Two years later, he left to continue his legal studies at the University of Leipzig. By then he had a tutor, Anders Sorenson Vedel, hardly older than Tycho himself and charged with keeping his nose in the law. Vedel must have been dismayed to find Tycho devoting so much of his time to astronomy. Yet Tycho was a headstrong and determined character; he worked at law during the day and stole out at night to study the stars while Vedel was asleep. By then he had acquired a celestial globe "no bigger than a fist" as well as star maps published by Albrecht Dürer by means of he came to recognize all the constellations. But as yet he had no proper instruments, and "could only check the predictions of the ephemerides by lining up a planet and two stars by means of a taut string and estimating the positions of the planet from the positions of the two stars on his little globe."[2]

The next turning point in Tycho's career occurred in August 1563, when he observed a conjunction of Jupiter and Saturn. Instead of a taut string he was now using a pair of large compasses, sighting from the vertex along each leg to the two

---

[1]Quoted in J.L.E. Dreyer, *Tycho Brahe: a picture of scientific life and work in the sixteenth century*. Edinburgh, Adam & Charles Black, 1890, p. 14.

[2]Victor E. Thoren, *The Lord of Uraniborg: a biography of Tycho Brahe*. Cambridge: Cambridge University Press, 1990, p. 16.

Albrecht Durer's star chart. Chicago Art Institute, photograph by William Sheehan, 2008

objects being observed. Even with these rudimentary means, the 16-year-old Tycho was able to establish that the *Alfonsine Tables*, based on the planetary theories of Ptolemy and drawn up at the order of King Alfonso X of Castile in the thirteenth century, were wrong by a month in their prediction of the date of the conjunction, while even the tables that replaced them, Erasmus Reinhold's *Prutenic Tables* based on Copernicus's theories and published only in 1551, were already in error by several days. Tycho then resolved to make his life work the construction of more accurate tables of the planetary motions which, he realized, meant obtaining better and more complete observations than any then in existence. At the age of only 16, writes his biographer J.L.E. Dreyer,

> his eyes were opened to the great fact, which seems to us so simple to grasp, but which had escaped the attention of all European astronomers before him, that only through a steadily pursued course of observations would it be possible to obtain a better insight into the motions of the planets, and decide which system of the world was the real one.[3]

Henceforth, the law could no longer hold him and when, soon after the conjunction, Joergen died, Tycho received through inheritance the financial independence he needed to pursue his interests. From Leipzig, he moved to Wittenberg and then to Rostock (where he famously lost his nose in a duel; it was fought not over a woman but a mathematical point over which Tycho and his fellow combatant had disagreed). Ever resourceful, Tycho promptly had the missing member replaced with an artificial nose of copper and silver. Some of his portraits also suggest that a branch of the facial nerve may have been severed; his face looks decidedly lopsided. Perhaps self-consciousness about his appearance explains why he would marry a commoner instead of a woman of noble extraction). He moved again, to Basel, then to Augsburg where he began to acquire proper scientific instruments suitable for a professional.

Whereas Copernicus had once told Rhaeticus that he would have been as pleased as Pythagoras was said to have been on discovering his theorem if he could make observations accurate to 10 min of arc, Tycho would aim at observations accurate to only 1 min – the ultimate attainable by a naked-eye observer. (At the time, there were no instruments available close to achieving this level of accuracy).

Tycho's first proper astronomical instrument was a cross-staff, in which the crossbar could be moved along the staff until the angle to be measured was exactly covered by the length of the bar. He was using this instrument by May 1564 but, though professionally made, it "simply would not perform to the expectations of a 17-year-old."[4] At Augsburg he began to design his own instruments. One consisted of a pair of giant compasses, with a 30° brass arc and wooden legs one and a half meters long; it was light and easily transported but had a systematic sighting error limiting its accuracy. He next devised for a wealthy colleague a large quadrant, consisting of a 90° arc made of well-seasoned oak to which were attached a brass graduation strip and plumb bob. The radius was five and a half meters and it was so heavy that 40 men were needed to put it in place.

The great quadrant was built in 1570. The next year Tycho left Germany and returned to Denmark. After a brief stay at the family estate at Knudstrup where he was soon bored with a life of "horses, dogs and luxury," he moved to Herre Vad, where he established a small private observatory on the estate of another well-to-do uncle, Steen Bille. Steen Bille alone among Tycho's relatives approved of his scientific tastes, and dabbled in alchemy himself.

This, then, was Tycho's situation when, on November 11, 1572, he noticed on returning from Steen Bille's alchemical laboratory to supper a brilliant star near the familiar chair of Cassiopeia. The star, standing nearly overhead, shone as bright as Venus, and

---

[3]Dreyer, *Tycho Brahe*, p. 27.

[4]Thoren, *Lord of Uraniborg*, p. 31.

was visible in broad daylight. Nothing like it could be recalled in living memory.[5] It remained conspicuous for a time, then began to fade, and after 18 months slipped below the threshold of naked-eye visibility. Tycho used his instruments to show conclusively that it was exceedingly remote, far beyond the sphere of the Moon – and not only farther but much farther, to all intents and purposes located in the sphere of the "fixed stars." A decisive blow had been struck against the Aristotelian cosmos, in which everything was in its right place in a divinely fashioned hierarchy. That blow had been struck by a Danish nobleman, who had stepped out of role in making astronomical observations, a task which, by the standards of the day, was more suitable for a tradesman.

Tycho published his observations in a book, *De Nova Stella*, which made him famous, and added to his fame in making a series of observations of the Great Comet of 1577, showing it to be at least six times more remote from the Earth than the Moon. He also proved that it should have passed through the crystalline spheres, if any existed, and thereby smashed another venerable Aristotelian conception.

For sometime Tycho had been considering moving back to Basel, but the Danish king, Frederik II, hoped to retain him in Denmark. Frederik dangled before him an offer too good to refuse: he would receive the island of Hven (a name Tycho always insisted meant "the island of Venus") in the sound between Kattegat and the Baltic Sea, "with all our and the Crown's tenants and servants who live thereon, with all rent and duty that comes from that, to have, enjoy, quit and free, without any rent, all the days of your life." Tycho also received the funds to build a splendid observatory and to pay for its upkeep. He was asked only to perform a few nominal duties, such as keeping the Cathedral at Roskilde in good repair.

Tycho accepted Frederik's generous offer, and so, above the white cliffs rising from the sea, he built Uraniborg, the "castle of the heavens." The architecture was baroque – the "castle" looked rather like a gingerbread house with an onion dome and cylindrical towers within which Tycho set up a gallery of instruments including sextants and quadrants, every one of which had open sights, since they were meant

[5]In fact, another such star had blazed in the constellation Taurus in 1054 and been recorded by Chinese astronomers and even, apparently, by Indians in the American Southwest, who seem to have set down their observation in a petroglyph on an overhanging rock ledge in New Mexico. Christian Europe, however, would seem to have taken no notice of the event in its chronicles. One had to go all the way back to Hipparcos to find anything like it in European records. In the second century BC, Pliny says in his *Natural History*, Hipparcos had seen a new star which had been presumed merely a very unusual and persistent kind of comet. Comets were then thought to be exhalations of some sort in the Earth's own atmosphere. Indeed, Aristotle had taught that everything changeable was confined to the region below the Moon while the heavens were a realm where all was eternal and unchanging. By establishing that the star of 1572 was well beyond the orbit of the Moon and indeed in the sphere of the fixed stars, Tycho showed that Aristotle's notion was unsustainable. Only 20 years before Robert Recorde, remembered as the first Englishman to embrace the Copernican theory, had written in *The Castle of Knowledge* (1551):

"Yea thoughe all other thinges in the worlde by tyme be consumed, and euen the moste harde metals freted into drosse, yet the liquide heauens not only gouerne time it selfe, but vtterly stande cleere from all corruption of time."

Tycho had shattered that romantic notion forever.

to be used with the naked-eye. He was a driven perfectionist; the same personality traits that later made him an arrogant and much-hated master for his subjects on Hven also made him the most exactingly fastidious observer the world had yet seen.

Tycho and his great mural quadrant. Handcolored engraving in *Astronomiae instauratae mechanica* (Wandesburg, 1598)

Before Tycho, problems of sighting with instruments had largely gone unrecognized, but no complication was too small for Tycho to consider if it stood in the way of his goal. To eliminate sighting errors he tried various expedients, eventually devising an apparatus with a metal plate having two horizontal slits, one above the other at a distance equal to the diameter of a cylindrical peg placed in the center. If, on looking alternately through the slits, the eye saw equal parts of the star project above and below the peg, then he knew the instrument was exactly pointed on the star. He also improved the accuracy of the divisions of his graduated circles by using transverse rows of points that allowed him to read the smallest subdivisions.

In addition to his astronomical studies, Tycho was devoutly attached to alchemy – recent analysis of the hair and hair-roots of his mustache have shown that he had mercury levels 100 times above normal – and regularly doled out medicines he concocted to his subjects. (He appears to have taken in an enormous amount of mercury shortly before his death; as to the significance of this, see below.) He was also as astrologically-minded as the Babylonians, and shared the reverence for the heavenly bodies of the old astronomer-priests: it is said he never observed except in his most luxurious robes. A seemingly "indestructible, blustering social being with an enormous appetite for food and wine," he was "fond of mystery and display, and his observatory at Uraniborg abounded in mechanical devices and imperceptible means of communication with which he liked to mystify his visitors.... Attached to the observatory was a dwarf called Jep, whom Tycho used to feed with an occasional morsel at table, like a dog. Jep was supposed to be clairvoyant and to have made some remarkable prophecies."[6] What dramas must have transpired on Hven in the Baltic, under the imperious majesty of the Lord of Uraniborg!

Tycho's mature – that is, his "virile, precise, and absolutely certain" – observations began in 1580. He set out to produce a star catalog more accurate than those of Hipparcos and Ptolemy, and he also carried out important studies of the motions of the Sun and Moon. But from the beginning his chief preoccupation was the planets, as was inevitable for someone with his astrological proclivities.

Mars – the planet most obstreperously defiant of following the path prescribed by eccentric and epicycle, orb in orb – especially drew his attention. He observed it carefully at every opposition from 1580 onwards, and not only at opposition. A Tychonic innovation – in the end it proved crucial – was that he also kept it under surveillance at other times.

In 1583, Tycho noted that near opposition Mars was moving retrograde at a rate of nearly a half degree every day. This seemed to prove that the planet could approach much nearer to the Earth than the Sun, something possible in the Copernican but not in the Ptolemaic system. Nevertheless, though almost alone among his contemporaries in not accepting Osiander's false preface as authentic, Tycho could not follow Copernicus all the way: the idea of the "heavy and sluggish Earth" moving through space seemed a stumbling-block, and he was even more distressed with the idea that if the Earth did follow an orbit around the Sun the stars would have to be

---

[6]Herbert Dingle, "Tycho Brahe," in *Astronomy*, edited by Samuel Rapport and Helen Wright, New York: Washington Square Press, 1964, p. 42.

both impossibly remote and improbably large.[7] In the end he devised the Tychonic system – a compromise between the Ptolemaic and Copernican systems in which the planets, excepting the Earth, traveled around the Sun while the Sun and Moon revolved about a stationary Earth.

The 1595 opposition of Mars was the last Tycho observed at Hven. After 20 years, the imperious lord of Uraniborg had worn out his welcome. He had pursued his exacting course of observations using instruments that he designed "with effort, attention to detail, and at unbelievable expense," and they were paid for by heavy taxation of his long-suffering tenants. Though he was selflessly devoted to astronomy, he failed to carry out even the lightest obligations the King had imposed on him in return for his privileges. When his patron Frederik died, the Danish court unceremoniously cut off his funds, and Tycho, in 1597, went into exile; leaving Hven forever, he took with him his treasure-trove of precious (but haphazardly organized) observations and a few of the most portable of his instruments. He left behind not only the observatory of Uraniborg but a second-generation observatory, Stjerneborg ("Star Castle"), built in 1584, in which the instruments had been located underground so as to be out of the way of wind and weather. Though he wrote a dedication for Stjerneborg in gold lettering on durable porphyry expressing his hope that his observatory with its precious instruments might not "totter through age nor any other misfortune, nor be transferred elsewhere, nor be violated in any manner," the observatory he had built for "all eternity" was, as Randall Rosenfeld quips, now "on the run."[8]

The observatories at Hven soon vanished without a trace. When his first biographer, Gassendi, in 1647 visited Hven in search of remnants of them, he found nothing. "There is in the island a field where Uraniborg was," he wrote. Tycho did record for posterity likenesses of the buildings and instruments – including his mural quadrant, equatorial armillary, and great celestial globe – in a lavish treatise published in 1598, in which "even now, some four centuries later, the likenesses of the instruments communicate a sheer physicality which can still impress."[9] Ultimately, however, his imperishable legacy was the observations he carried with him.

The noble refugee arrived first in Germany, taking up temporary residence at Wandsbeck Castle. His reputation preceded him, for at the very moment he was

---

[7] Tycho could discern no parallax when he observed the stars at points which would correspond to alternate sides of the Earth's orbit if the Copernican theory were true. This meant that they would be extremely remote. At the same time, he believed (as the Arab astronomers had done) that the stars showed small disks. According to Tycho's figures, first magnitude stars subtended an angle of 120″; second 90″, third 45″, fifth 30″, sixth 20″. Dreyer notes in *A History of Astronomy from Thales to Kepler*, p. 361: "Now if the annual parallax of a star of the third magnitude was as great as one minute, the star would be as large as the annual orbit of the Earth around the Sun. And how big would the brightest stars have to be, and how enormously large would they be, if the annual parallax was still smaller?" Only with the invention of the telescope did this line of reasoning cease to carry force.

[8] R. A. Rosenfeld, "From Uraniborg to Yerkes: a fragile monumentality." *Griffith Observer*, February 2009, 2–18:5.

[9] Ibid., p. 3.

working on the lavish treatise just described, a book arrived from a young German mathematician. It was an eccentric work, but showed ability. The name of the author, Johannes Kepler, was unknown to him. He can hardly have imagined then that Kepler's name and his would become forever linked.

Johannes Kepler. One of the few portraits of him known to have been painted during his lifetime, and showing him at the height of his powers. In 1620, Kepler sent it to his friend Bernegger in Strasburg by his assistant Gringalletus. It is now in the Thomas seminary in Strasburg

Kepler had been born in 1571 at Weil-der-Stadt, a village in Swabia, in southern Germany. He was a scrupulously devout Lutheran – since boyhood, says his biographer Max Caspar, his "religious disposition was pronounced."[10] At the age of four, he was smitten by smallpox; it nearly killed him, and left him with a frail constitution and with his eyesight permanently damaged. As a result, he would play no part in the observations with the telescope soon to revolutionize astronomy. Such observations were, he said, for him a "forbidden fruit." He went to a local school but when his father, Heinrich, a ne'er-do-well, went into debt, Kepler was recalled and employed

[10]Max Caspar, *Kepler*, trans. C. Doris Hellman. New York: Dover Publications, 1993, p. 40.

as a pot-boy in his father's tavern. His father – an immoral, rough, quarrelsome old soldier, as Kepler himself said – eventually abandoned his family, and though what became of him is unknown, it is hard to imagine he did not turn out badly. Meanwhile, Kepler's mother, Katharina, seems to have extremely ill-tempered; for all we know, she may well have resembled the popular idea of a witch, and in later years she was found guilty of witchcraft and condemned to imprisonment and torture. Kepler's personal intervention alone saved her from the fate of so many women during that period when witch hunting was reaching its peak in Germany – with no fewer than 38 in Weil-der-Stadt alone put to death in the years between 1615 and 1629. Into this dark and dangerous age Kepler would attempt to shine the light of his genius.

Despite the disadvantages he struggled against, Kepler's ability in scholarship shone through from the first. He was sent to a monastic school at Marlbronn and passed, at 17, to the University of Tübingen. In that religiously obsessed age, he was sure he had a religious vocation, and he began to prepare himself for a career in the Lutheran Church. However, at Tübingen he also fell under the magnetic spell of Michael Maestlin, an outspoken Copernican. Under Maestlin, Kepler became a passionate convert to what would become a new religion to him – Copernicanism – whose doctrines he was soon defending with many ingenious and elegant arguments.

Kepler remained a devout Lutheran to the end of his life and still hoped for a career in the Church when, in 1593, a position as a professor of mathematics at the University of Graz opened up. Malleable to circumstance, he accepted, though making it clear that he intended to leave whenever anything better opened up. The quality of students at Graz seems to have been rather poor; he had only a few students, and they must have found him difficult to follow. By all accounts he was an inspired – but not an inspiring – teacher.[11] His biographer Max Caspar tells us:

> If he found few listeners, then the fault was certainly in part his. He expected too much of his pupils and assumed they would have the same enthusiasm for his subject and the same devotion to the knowledge of truth by which he himself was animated. In a penetrating self-characterization which he wrote in 1597, he cited qualities which also throw light on his teaching. He speaks ... about his overpowering "*cupiditas speculandi*" [eagerness to speculate], of his philosophical urge which throws itself at everything and always tackles new things, which rushes on and robs him of leisure to think a thought through to the end. He would always think of something to say before he had a chance to reflect on how good it might be. So he always talked too quickly. He would always think of new words, new subjects, new modes of expression and new arguments while talking and writing. Sometimes in the middle of a lecture he would think that it might be better to change its purpose or even to suppress what he was saying.... Since his memory brought back all related subjects at the same time and as everything occurred to him at once, he also wanted to say everything at once.... Neglecting his very praiseworthy profession, he wandered off wherever his mind drove him...[12]

[11] Sharing this trait in common with Newton and Einstein, among others. See Ioan James, "Singular scientists," *Journal of the Royal Society of Medicine*, 96 (January 2003), 36–39.

[12] Caspar, *Kepler*, pp. 57–58.

These same qualities are manifested in his writings, and no doubt this helps account for why the significance of his results was not at first appreciated by his contemporaries. His first – and, in his own opinion, his greatest – discovery occurred, in typical Keplerian fashion, with a flash of insight in the middle of a lecture he was giving at Graz in July 1595. He was explaining to his students how conjunctions of Jupiter and Saturn – their linings up on the same side of their orbits relative to the Sun, such as Tycho had observed as a student in August 1563 – take place at successive points around the Zodiac separated by some 240°. To make his point clearer, he drew one circle to represent the Zodiac on which he marked off the points of the successive conjunctions and connected these points with a series of line segments. In this way he sketched a series of triangles whose overlapping figures defined a circle standing in relation to the greater circle of the Zodiac in about the same proportion as the orbit of Jupiter stood to that of Saturn. Encouraged by this success, he placed a square inside the orbit of Jupiter, in hopes that the inscribed circle would be of the right dimensions to correspond to the orbit of Mars. It didn't work. But undaunted, he attempted a different solution: in fact, it was a mere throw of the die. Why might not the orbits of the planets be defined by a series of inscribed and circumscribed spheres constructed around a set of polyhedra?[13] At once he knew the identity of the polyhedra concerned: "Anyone with the least knowledge of geometry, would immediately think of the five regular solids and their related circumscribed and inscribed spheres." These were the so-called Platonic solids that, Kepler remarks, had been "celebrated from the time of Pythagoras and Plato down to our own." Euclid had proved that there are five and only five of these regular solids: the Cube, the Tetrahedron, the Dodacahedon, the Icosahedron, and the Octahedron. There were five, neither more nor less. Only five.

Now Kepler, at the age of 24, believed he had discovered nothing less than the secret of the universe's structure! (One can only sympathize with his students at that lecture in July 1595. They would hardly have been able to follow him *here*). There were only five regular solids; so Kepler thought that he could explain why there were six planets – Mercury, Venus, Earth, Mars, Jupiter, and Saturn – neither more nor less. At least his reasons were (slightly) better than those of the Fool in King Lear: "Dost thou know why there are only Seven Sisters? Because they are not eight." He also was able to compute – a priori, by deduction alone – the relative distances of the planets from the Sun, which he compared with those Copernicus had published in *de Revolutionibus*. At least the results were not disastrously off base, and in any case the discrepancies did not bother him much, since Copernicus, as he knew, had referred the planetary motions not to the Sun which he had made the center of the universe but to the center of the Earth's sphere (a concession to the fact that the observed motions of the heavens seem to us, as observers on the Earth's surface, to be taking place with reference to the Earth). Kepler expected that, once these values were recalculated (which he undertook to do with Maestlin's assistance), he would see better agreement.

[13]At this stage, Kepler was still thinking in terms of spheres, as Copernicus had done also; his book, after all, had been called the "Revolutions of the Heavenly Spheres."

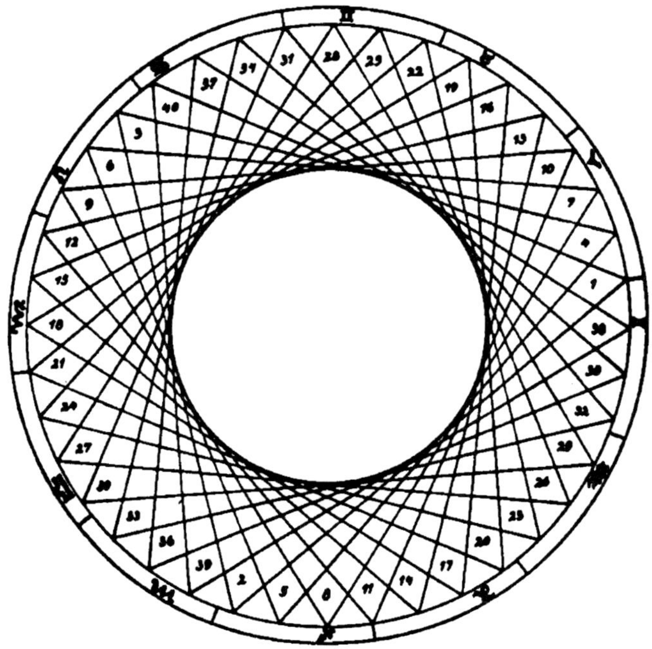

Kepler diagram showing the progression of the conjunctions of Jupiter and Saturn, the germ of the idea he developed in *Mysterium Cosmographicum*

Certainly he was convinced he had made a great discovery, and wrote of the intense pleasure it gave him. It was such as could "never be told in words. I regretted no more the time wasted; I tired of no labor; I shunned no toil of reckoning, days and nights spent in calculation, until I could see whether my hypothesis would agree with the orbits of Copernicus or whether my joy was to vanish in the air."[14]

Kepler wrote a book describing his theory, *Mysterium Cosmographicum*, and it was this he sent to Tycho at Wandsbeck Castle. Kepler – proud parent that he was – expected his literary progeny to receive the adulation of the astronomical world. Instead it hit with a marked thud. Tycho was more encouraging than he needed to be. He had his doubts but conjured up something that sounded vaguely like encouragement. "Do not build up abstract speculations concerning the system of the world," he told his younger colleague, "but rather first lay a solid foundation in observations…."[15] In general Tycho – despite his well-earned reputation as a difficult personality – acquitted himself admirably in his relations with Kepler. He was careful not to be unduly dismissive. For that matter, he had need of such a man. From Wandsbeck Castle he sent a letter to Kepler praising him for the "ardor you have shown in making

[14]Ivor B. Hart, "Johannes Kepler," in: *Astronomy*, ed. Rapport and Wright, p. 45.

[15]Ibid., p. 46.

these enquiries." Though he did not think that Kepler was right in everything, he did believe that "by using the true values of the eccentricities of the planets as obtained by myself over many years, it would be possible to make more accurate verification" of Kepler's theories. Just then he was too busy to offer more material assistance; however, he indicated that perhaps it would be possible for him to do so later.

As mail traveled slowly in those days, Kepler did not receive Tycho's letter until February 18, 1599. He could not make the journey to Wandsbeck; it was too far, too difficult, and too expensive (Kepler had a wife and three children to support by then). Moreover – always thin-skinned – he had been somewhat injured even by Tycho's mild criticisms. Nevertheless, he did not forget Tycho's invitation, and at the end of 1599 learned that Tycho had been appointed "Imperial Mathematician" by the Holy Roman Emperor, Rudolph II. Benateky Castle, on the river Yser about 22 miles northeast of Prague, was placed at his disposal, and there Tycho, gathering up the effects of a ruined career, with an assistant Christian Severinus (who called himself Longomontanus) worked desperately against time to reduce the observations he had hoarded at Hven.

He was in need of another calculator. His invitations to Kepler were sincere, and by 1600, Kepler was also a refugee. After mulling things over, at last he accepted Tycho's invitation to join him in Prague and assist him in the reduction of his planetary observations. Kepler (with his wife and children) arrived to find Tycho and Longomontanus hard at work on the theory of the motions of Mars. Kepler would be pressed to join them in the monumental task.

He later saw it as the work of Divine Providence. In summing up the strange circumstances that led him to concern himself with the theory of Mars, he noted that – still filled with ardor for the (supposed) discovery he had announced in the *Mysterium Cosmographicum* – he had first hoped to learn from Tycho whether his discovery could withstand comparison with accurate observations. However, "the distance between us" – Tycho was then at Wandsbeck and he at Graz – prevented him from doing so. When Tycho removed to Prague, he became more accessible. So, said Kepler: "I am firmly of the opinion that it was by the force of divine Providence that he came to Bohemia. I went to him then at the beginning of 1600 in the hope of being informed about the corrected eccentricities of planetary orbits."[16]

In the result, Kepler remained with Tycho for only 4 months. From the outset, relations between the two men were strained; Tycho's household was crowded and disorderly, and in addition Tycho – social, expansive, imperious and domineering as ever – could not have presented a stronger contrast with the reclusive and introverted scholar. Tycho, moreover, had no intention of imparting to Kepler any information he did not need to know for the computations he was assigned. He did not even allow him to use his instruments (in the end they would be scattered and destroyed – every last one – by war or fire!). Kepler's nerves were already shattered; he became depressed, and was often physically ill. Finally he could stand it no more.

[16] Alexandre Koyré, *The Astronomical Revolution: Copernicus – Kepler – Borelli*. Trans. R.E.W. Maddison. New York: Dover Publications, 1992, p. 164.

And yet though Tycho was notoriously difficult to get along with, Kepler admitted that he was more to blame for their falling out; he behaved, he said, like a "mad dog." He left Prague – as he then expected, forever. Returning to Graz, he hoped to resume his position at the Protestant Seminary, but when he arrived, he found it in the process of being closed down. In near-despair, Kepler briefly considered studying to become a physician – a desperate expedient indeed! (If he had done so, it is unlikely that medicine would have been much advanced; progress in astronomy would, however, have been set back at least a century.)

At last, setting his face like flint, he resolved to return to Tycho in Prague. "If God is concerned with astronomy," he wrote, "which piety desires me to believe, then I hope I shall achieve something … for I see how God let me be bound with Tycho through an unalterable fate and did not let me be separated from him by the most oppressive hardships."[17]

By now Tycho had hardly 2 months to live. He died on October 24, 1601, 11 days after becoming seriously ill during a banquet. According to the first-hand testimony of Kepler himself, Tycho had attended a banquet, at which he had too much to drink, and was either too polite (or proud) to relieve himself. When he returned home, says Kepler, he could not urinate any more. Uninterrupted insomnia followed; then fever, and finally delirium. During his final delirious night, "like a composer creating a song," Tycho repeated over and over again: *Ne frustra vixisse videar*, "Let me not seem to have lived in vain."

The cause of Tycho's death – complications of a bladder obstruction – has lately been called into doubt. There have even been suggestions that Tycho was poisoned – either by Kepler (implausible, despite Kepler's interest in getting hold of Tycho's observations; for one thing, Tycho was at the time negotiating with Rudolf to get Kepler appointed Imperial Mathematician, so it would have been completely counterproductive to kill him!) or by an assassin employed by King Christian IV of Denmark who supposedly suspected Tycho of having an affair with his mother.[18] In all likelihood, Tycho himself, in the agony of his dying illness, overdosed on his own mercurial compounds. In any case, after having amassed 38 years of observations, Tycho's life was over – but his work was not yet done. With his arrogant, disturbingly lopsided face and walrus mustache, Tycho would have been well cast as another character with a charge to the next generation, as Hamlet Senior's ghost in Shakespeare's play, just then premiering at the Globe, while Kepler was a reasonable casting for Hamlet himself. Tycho died in the hope that his observations would be used to establish his Tychonic (not the Copernican) system; his final utterance

[17]Ibid., p. 123.

[18]On Kepler, see: Joshua Gilder and Anne-Lee Gilder. *Heavenly Intrigues: Johannes Kepler, Tycho Brahe, and the Murder behind one of history's greatest scientific discoveries*. New York: Anchor, 2005. On Christian IV, see: Matthias Schulz, "Was Tycho Brahe Murdered by a Contract Killer?" *Spiegel Online International* (October 5, 2009).

Tycho's effigy over his burial place in Tyne church, Prague. Photograph by Richard McKim

was not unlike Hamlet, Senior's admonition of "Remember me!" to Hamlet. Poor Kepler was the one "born to set it right." He might well have said, with Hamlet,

> ... Remember thee!
> Ay, thou poor ghost, whiles memory holds a seat
> In this distracted globe. Remember thee!
> Yea, from the table of my memory
> I'll wipe away all trivial fond records,
> All saws of books, all forms, all pressures past
> That youth and observation copied there,
> And thy commandment all alone shall live
> Within the book and volume of my brain,
> Unmix'd with baser matter. Yes, by heaven!
> (*Hamlet*, I, v)

The 1600s were marked by prodigies, both natural and human. In February and March of that year, a volcano – Huanyaputina, located 70 km east of Arequipa, Peru – erupted in a tremendous blast comparable to the celebrated eruptions of Tabora in 1815, Krakatoa in 1883, and Pinatubo in 1991. As far away as Europe the Sun and Moon were dimmed and reddened by ash, and Europe suffered from the coldest decade in centuries. A New Star – like the one Tycho had seen in Cassiopeia – erupted in Ophiuchus in 1604. Bruno was burned at the stake in Rome, Shakespeare finished *Hamlet* and then (in quick succession) *Othello*, *Macbeth*, *King Lear*. Kepler received the torch from the hand of the dying Tycho. (It is not always realized, by the way, that Shakespeare's greatest works – and Kepler's, and Galileo's – occurred in the same crowded decade, the first decade of the 1600s.)

After Tycho's death Kepler (not without difficulty) managed to wrest control of the observations of Mars from Tycho's heirs. They were scattered through many pages of Tycho's notebooks and as disorganized as Tycho's Bohemian household had been. Kepler's task was to bring order into the confusion and use the observations

Monument to Tycho and Kepler in Prague. Photograph by Richard McKim

of Mars to improve the theory of the motions of that planet – and so to honor Tycho's dying wish "not to have seemed to live in vain."

Kepler succeeded Tycho as Rudolph's Imperial Mathematician, and moved into rooms in Prague itself, so he could be closer to Rudolph.

Rudolph himself cut a curious figure. He was a Habsburg, a great-grandson of the Habsburg emperor Charles V, the last to attempt to realize the fading medieval dream of a spiritually and politically unified Christendom. Rudolph had acceded to the throne of Holy Roman Emperor in 1576 but was completely miscast as ruler; shy, anxiety-ridden, unsociable – traits that only increased with age – he never married, and holed up in his castle in Prague. He preferred intellectual pursuits to the cares of statecraft. At times, he was simply unable to face the world. He would shut himself for days at a time in his "treasure room," where he could devote himself without interruption to collections of paintings and sculptures, textiles, gems and coins, mechanical clocks and toys, alchemical apparatus, wax figures, monsters. It was perhaps but an extension of this madcap character's interests that he also collected the great men of Europe, as many as he could; sometimes, as in Tycho's case, he did so by offering them exorbitant sums, though inevitably he found it impossible to pay them. Drawn to alchemy, he had a cadre of alchemists working for him in the Castle, and probably absorbed large amounts of mercury into his system (he did exhibit signs of mercury poisoning, later called "mad-hatterism"). He was no less keen about astrology. Kepler was not opposed to drawing up a horoscope for him now and again, and realizing Rudolph's vulnerability did what he could to prevent unscrupulous persons from preying on his worried and darkly credulous mind. Perhaps he saw that Rudolph was often tottering on the verge of insanity. For all his faults, at least Rudolph did not share the fanaticism about religion that was the scourge of the age. Though devoutly Catholic, he did not insist that the scholars who came to him profess his faith. Apparently he was not all that interested. Late in his reign he went so far as to publish a letter granting all Bohemians against the forceful imposition of religion, a generous gesture that unfortunately had little effect in an age when fanatic intolerance trumped tolerant moderation.

Down the hill from the castle in which Rudolph tried as best he could to shutter himself from the cares of empire, Kepler calculated. He pored over Tycho's observations and tried to squeeze from them the shape of the Martian orbit. He first established that Mars's positions in its orbit always fell slightly inside where they would have been if its orbit were circular. He next tried various ovals, and confided to a friend, David Fabricius, just how easy things would be were the orbit the mathematically simplest oval, the ellipse, whose properties had already been worked out by Apollonius in the third century BC. He found a short-cut in his calculations when he discovered that the line connecting Mars with the Sun always swept out equal areas in equal times; in other words, Mars travels fastest when it is nearest the Sun, slowest when it is farthest away – a relation that, though the first discovered, is now known as Kepler's second law. At last, at Easter 1605, he experienced another *eureka*! moment. It dawned on him that the orbit of Mars truly is an ellipse, with the Sun at one focus. There have been, in the history of astronomy, a mere handful of discoveries as momentous as this. After such a discovery, the rest of Kepler's life might well have been malingering. That it wasn't is a great tribute to Kepler's drive and badger-like tenacity.

The *Astronomia Nova* in which he announced his immortal result is – in marked contrast to Ptolemy's *Almagest* or Copernicus's *de Revolutionibus* – wonderfully engaging, ornately poetic, full of personal information about what Kepler felt as he followed up leads and encountered disheartening set-backs. It describes his bewildering trek through months of frustration and near-despair redeemed by sudden lashes of insight and elation. These qualities make it one of the most useful sources of the psychological process and the unconscious sources of inspiration of scientific discovery that we have.[19]

The rest of Kepler's life was full of trials. He applied for funds to extend his work beyond Mars, and from his rooms in Prague laid plans to extend his siege to the other planets.

Poor Rudolph was always chronically short of funds; he did not have the money to fight all his battles on Earth, much less to wage planetary warfare! Even the money to publish Kepler's book could not be found immediately; thus the delay in its appearance from 1605 until 1609. Kepler's salary was continually in arrears. He was often obliged to appear at Court and to suffer the humiliation of groveling to be paid. At the end of 1610 his first wife, who seems to have been prone to depression and was probably difficult to live with, came down with fever, suffered seizures, and died. His 6-year-old son – his favorite – succumbed to smallpox. It was a terrible year and yet – despite his personal travails – he absorbed at once the important discoveries Galileo had announced with the telescope.

By 1612, Rudolph himself was dead, and with Prague itself about to become a battleground in the Thirty Years' War, he fled to Linz, Austria, where – despite the unhappiness of his first marriage – he immediately began casting around for another wife. No less than 11 candidates were considered, and Kepler – setting forth the merits and demerits of each with the detachment of a mathematician – chose the poorest, an orphan girl without a dowry. This marriage proved to be much happier than the first, and his wife promptly bore him seven children (apparently there are descendants of one of Kepler's daughters living to this day). These abundant progeny were a blessing, but they can hardly have eased the strain on his finances. Also during these fatiguing years Kepler traveled to Würtemberg to defend his mother against the charge of witchcraft.

In spite of all, he worked. Through years of ill-health, poverty, worry, and incessant labor, he persisted; his studies – despite wife, children, financial difficulties, the world careening toward chaos – must have long since become for him a "refuge as well as a passion."

In Linz in May 1618 – in a house that still survives – he made yet another magnificent discovery. The harmony of the world had haunted him ever since the

[19] Fortunately, it is available in a superb English translation: Johannes Kepler, *New Astronomy*, trans. William H. Donahue (Cambridge: Cambridge University Press, 1992).

early days in Graz when he had believed himself to have grasped, in a blinding flash of insight, the mystery of the universe. Even after he joined forces with Tycho and realized that his initial formulation was in need of modification, he hoped to unravel the relationship he felt existed between the distances of the planets and their average speeds around the Sun. For two decades, the relationship eluded him.

One can picture Kepler laboring over his desk: a small, delicate figure, at least one admirer found it difficult to comprehend how "such a mass of learning, and knowledge of the most profound secrets, could be locked and concealed in one such small body."[20] He described himself as small, active, but gnarled; nearsighted and suffering from multiple images in one eye – a condition not helped by his habit of incessant night study; prone to bilious attacks from any diet except the blandest, and used to gnawing at bones and dry bread; subject to boils and rashes; averse even to baths and washings. His mind and body existed in a perpetual state of warfare. As a youth, he had compared his nature to that of a "little house dog," and he always had a dependent and subordinate nature; yet because of his sincere piety, he seems to have felt all the more the leadings of a higher hand that guided his destiny. Nevertheless, Caspar says, "between his genius and his humanness there remained a latent gap. Although the idea of harmony kept his thoughts busy, he was not harmonic, not adjusted in his nature. He was a restless soul, fluctuating repeatedly between exhilaration and depression."[21]

Now, after many years of depression, he experienced one more supreme moment of exhilaration. It is best to let him describe in his own words:

> What I prophesied two and twenty years ago, when I discovered the five solids among the heavenly orbits … that for which I joined with Tycho Brahe, for which I settled in Prague and for which I have devoted the best part of my life to astronomical contemplations, this at length I have brought to light. It is not eighteen months since the first glimpse of light reached me, three months since the dawn, very few days since the unveiled sun, most admirable to gaze upon, burst out upon me. Nothing holds me; I will indulge in my sacred fury; I will triumph over mankind by the honest confession that I have stolen the golden vases of the Egyptians to build up a tabernacle for my God, far away from the confines of Egypt. If you forgive me, I rejoice; if you are angry, I can bear it; the die is cast; the book is written; to be read now or by posterity, I care not which; it may well wait a century for a reader, as God has waited 6,000 years for an observer.
>
> If you would know the precise moment, the idea first came across me on the 8th of March of this year, 1618; but, chancing to make a mistake in the calculation, I then rejected it as false. I returned to it again with new force on the 15th of May; and it has dissipated the darkness of my mind, by such an agreement between this idea and my seventeen years' labor on *Brahe's Observations*, that at first I thought I must be dreaming, and had taken my result for granted in my first assumptions. But the fact is certain, *that the proportion existing between the periodic times of any two planets is exactly the sesquiplicate proportion of the mean distances of the orbits*.[22]

---

[20] Caspar, *Kepler*, p. 369

[21] Ibid., pp. 371–372.

[22] Hart, "Johannes Kepler," pp. 52–53.

As we would now say: the squares of the periods of revolution of the planets are as the cubes of the mean distances. Here is the relationship that would eventually disclose to Newton – and others – the inverse-square law of gravitation. It was the crowning achievement of Kepler's life.

Having seized the wandering planets in the net of his calculations, Kepler worked tirelessly on his *Rudolphine Tables* – ephemerides of the planetary motions he had promised the long-dead Tycho he would complete. The work of computing them was backbreaking enough, but on top of all that, Kepler also had to try to scrape up the funds to publish them. Nor were his perennial peregrinations yet at an end; his misfortune – no fault of his – was to have been born into a violent and convulsive age. In 1618, the Thirty Years' war broke out, and in 1626 Linz itself came under siege. Kepler had to flee again, eventually finding refuge at the court of the General-in-Chief of the armies of the Holy Roman Empire, Albrecht von Wallenstein (whose horoscope he cast) at his newly formed Duchy of Sagan, in Silesia. At last, in 1627, Kepler published the *Rudolphine Tables*.

Kepler, never one to spare himself and nothing if not loyal, worked out the motions in triplicate: not only according to the Copernican Theory but also according to the Ptolemaic and Tychonic. So in the end he made good his debt to Tycho. He was now nearly spent, his health broken. When in October 1630 he set out from Sagan to Regensburg in hopes of conferring with the then-emperor about yet another possible residence – and about getting paid what was owed him – the trip proved too much for him. He suffered a chill and after a short illness, on November 15, 1630, he died. He had already written his own epitaph: "Once I measured the heavens, now I measure the Earth's shadows." He was buried in the Protestant cemetery of St. Peter in Regensburg. Four years later, the whole churchyard was ploughed up by Bavarian and Imperial troops, and Kepler's bones were scattered.

Among the great figures of the Astronomical Revolution, Copernicus's resting place has now been identified in the Cathedral of Frombork; Tycho lies in Tyn Church in Prague; Galileo in Santa Croce in Florence; Newton in Westminster Abbey in London. Only Kepler's resting place has vanished. Yet in some ways it is fitting that it should be so of this man who was never quite of this world, one whose spirit was so often at odds with his body. In the end, Kepler fervently believed that it was not the body that mattered but the soul, "the incorporeal image of God."

If Kepler's soul is immortal, it is because it lives on in his Three Laws of Planetary Motion.

Chapter 8
Moon Over Padua

... the Moon, whose Orb
Through Optic Glass the Tuscan Artist views
At Ev'ning from the top of Fesole,
Or in Valdarno, to descry new Lands,
Rivers or Mountains in her Spotty Globe.

Milton, Paradise Lost, I, 287–291

In the autumn of 1608, a spectacle-maker in Flanders named Hans Lipperhey (or Lippershey) applied for a patent for his invention of an instrument "by means of which visible objects, though very distant from the eye of the observer, were distinctly seen as if nearby." The States General in The Hague had hardly commenced considering this application when two other claimants stepped forward. In the end, Lipperhey received payment for his device but was denied the patent on the grounds that the principle of it was already too well known. Indeed, word spread quickly. By the following spring spyglasses were being offered for sale in the shops of spectacle-makers in Paris, and Galileo Galilei, professor of mathematics at the University of Padua in the "Serene Republic of Venice," learned of it from a former pupil, Jacques Badovere. Realizing that the device was based on the principle of refraction, Galileo at once fitted two lenses – one plano-convex to serve as the objective, the other plano-concave to serve as the eyepiece – at either end of a lead tube, according to the legend made from the sawed-off pipe of a church organ. This first instrument magnified 3×, only about the same magnification as a good opera glass.

An enterprising spirit whose first and so far only book was a manual on how to use a military compass he had invented as a means of supplementing his salary at the University, Galileo must have dreamed that his ship had finally come in. If he could improve this instrument enough, he might make a good deal of money. By the end of the summer he had managed to step up the magnification to 8×. Now he had an instrument powerful enough to interest the traders on the Rialto who eagerly awaited – like Antonio in Shakespeare's *Merchant of Venice* – the arrival of their wealth-laden "argosies with portly sail." For them an edge over other speculators could mean fortunes. Before long he had demonstrated it to the Doge and Senators from highest campaniles, including, one presumes, the Tower of San Marco.

W. Sheehan, *A Passion for the Planets: Envisioning Other Worlds, From the Pleistocene to the Age of the Telescope*, DOI 10.1007/978-1-4419-5971-3_8,

Galileo as he appeared near the end of his life. Painting by Justus Sustermans in 1636. By then he was blind in one eye, and his sight in the other was failing; nevertheless, he succeeded in making one more discovery – that of the librations of the Moon. Uffizi Gallery, Florence. Photograph by William Sheehan, 2008

Galileo was then 45, at the height of his powers if not yet of his fame. A feisty Italian with flaming red hair, he was full of fire and passion, a man bubbling with ideas. Vain, argumentative, possessed of rapier-like wit and keenly aware of his intellectual superiority, he was not one to suffer fools gladly – something that was bound eventually to get him in trouble. Though unconventional in a way, he was also very much a child of his time: eager for money and glory, and determined to climb the social ladder of his time.

He had been born on February 15, 1564 (2 months before Shakespeare) in Pisa, then ruled by the Grand Dukes of Tuscany. The roots of his family were in Florence, where the Italian Renaissance had begun and whose remarkably enlightened – and at first unostentatious – civic leaders, the Medici, had bankrolled it. The founding figure had been Lorenzo the Magnificent (1469–1492) who, in calculating the amount of money the family had spent on artists, architects and writers, found it "unbelievable"; he nevertheless added, characteristically, that it was "money well spent."

The Tower of San Marco, from which (in all probability) Galileo demonstrated the telescope during the summer of 1609. Painting by Canaletto, in the Chicago Institute of Art. Photograph by William Sheehan, 2008

Galileo's father, Vincenzio, was a Florentine. He was a musician – an accomplished lutist, composer, and a writer on music theory when the subject was still treated as a specialized branch of mathematics. Like Leonardo, Vincenzio believed in experience over authority. Defying his own teachers by tuning his lute not according to the abstract notions of the Pythagorean theory of harmony but according to his own estimation of the sweetness of sounds registered by his ear, he would write in *Dialogue of Ancient and Modern Music* (printed in Venice in 1581):

> It appears to me that they who in proof of any assertion rely simply on the weight of authority, without adducing any argument in support of it, act very absurdly. I, on the contrary, wish to be allowed freely to question and freely to answer you without any sort of adulation, as well becomes those who are in search of the truth.[1]

This passage could have been written 30 years later by his son.

Though Vincenzio returned to Florence in 1572, Galileo and his younger siblings lingered in Pisa in the care of relatives for two more years. In 1574 – the year that the first Grand Duke of Tuscany, Cosimo I de'Medici, died – Galileo was sent to continue his education at the picturesque Benedictine Monastery at Vallombrosa – the "Shady Vale." There he was given instruction in Latin, Greek, logic and religion and, much to his father's alarm, flirted with the idea of becoming a priest. Families were then required to provide the means needed to support children

[1]Quoted in Dava Sobel, *Galileo's Daughter*, p. 17.

in holy orders, and Vincenzio, never without money worries himself, had other, more lucrative ambitions for his gifted eldest son. He removed him from the Monastery, ostensibly because he was suffering from eye troubles requiring the attentions of a doctor. Possibly this was even true, as Galileo did suffer all his life from ailments of the eyes. At the end of his life he became completely blind, (though not, as is sometimes claimed, because he had damaged his eyesight by peering with his telescopes too steadily at the Sun).

Now Vincenzio took over Galileo's education at home before sending him to the University of Pisa in 1581. He always hoped that Galileo might become a physician. Attempting to oblige, Galileo did start out as a medical student, but proved a rather distractible one. While attending Mass one Sunday in the Cathedral, his attention was diverted to a swinging lamp. For centuries lamps had swung in the Cathedral. So far as we know, no one had paid any attention. But this swinging lamp seized Galileo's imagination, and started him meditating on the laws of pendulum motion. About this time he wandered into a lecture on geometry, and – with the enthusiasm of youth – at once became a devotee of the study of mathematics. Soon afterward he ran out of funds. Leaving the University without taking a degree – and lost forever to the study of medicine – he returned to his father's house in Florence in 1585.

Once back in Florence, Galileo took formal lessons in mathematics, and immersed himself in "geometric art," a term referring to subjects such as perspective and chiaroscuro (the technique for rendering depth through shadows and light). His teacher was the distinguished Ostilio Ricci, Mathematician to the Tuscan Court under Cosimo I's successor, Ferdinand II de'Medici. A fellow student was Lodovico Cardi, known as Cigoli, who became one of Galileo's closest friends and went on to achieve a high rank among artists of the day.

During this period Galileo busied himself, says his pupil and biographer Viviani, "with great delight and marvelous success in the art of drawing, in which he had such great genius and talent that he would later tell his friends that if he had possessed the power of choosing his own profession at that age … he would absolutely have chosen painting."[2] Clearly Florence – a city that was "extraordinarily self-conscious and proud of its great artistic tradition"[3] – had cast its spell over him. The few drawings by Galileo that survive from this time include sketches on two pages forming a protective binding for Galileo's student examination on Aristotle's *De coelo*, of which the head of a male figure "is not unskillful in its effect," while the face of a nymph "reveals practice." Clearly, Galileo was at least a competent artist. Horst Bredekamp, one of a number of scholars to consider the crucial subject of Galileo's relationship to the arts, concludes "the sketches bespeak a certain familiarity with the qualitative level of sixteenth-century drawing."

---

[2] Vincenzio Viviani. "Racconto istorico della vita del. Sig. Galileo Galilei." In *Opere di Galileo* (Florence: Barbera Editrice, 1890–1909), xix, 597–646:602.

[3] Samuel Y. Edgerton, Jr. Galileo, Florentine 'Disegno,' and the 'Strange Spottednesse' of the Moon, *Art Journal*, Fall 1984, 225–232:225.

Florentine vista from lantern of Duomo, showing the hills of Arcetri (*left*) and Bellosgardo (*right*). Photograph by William Sheehan, 2008

However, he qualifies: "To be sure, the *disegni* of the young Galileo are by no means masterpieces, and one could argue that it was best for Galileo that he did not follow a career as an artist."[4] But he did not forget what he learned, and one day his skill would allow him to master the subtle features revealed in the Moon.

Among Galileo's other activities in Florence, we know that he gave two public lectures on the inverted layered-cake cosmology of hell described in Dante's *Inferno*. More interestingly, he may have assisted his father in musical experiments, as Dava Sobel suggests:

> … when Vincenzio filled a room with weighted strings of varying lengths, diameters, and tensions to test certain harmonic ideas, Galileo joined him as his assistant. It seems safe to say that Galileo, who gets credit for being the father of experimental physics, may have learned the rudiments and value of experimentation from his own father's efforts…[5]

Galileo's Florentine interlude ended when he was recalled to Pisa to take up the chair of mathematics. Mathematics was hardly a prestigious field, and Galileo's

[4]Horst Bredekamp, "Gazing Hands and Blind Spots: Galileo as Draftsman," in: *Galileo in Context*, edited by Jurgen Renn. Cambridge: Cambridge University Press, 2001, 153–192:167.

[5]Sobel, *Galileo's Daughter*, p. 18.

salary was much less than that of the Philosophers who still taught chapter and verse from Aristotle. Moreover, he was never popular there. An iconoclast, he preferred to pay fines rather than wear the formal academic gowns required of the professors, and gave in to the temptation of writing trenchant, sometimes racy, verses lampooning the other professors. Instead of studying nature from ancient texts, he performed experiments, and recorded the results in writings never published during his lifetime. In the best known of them, he dropped unequal weights from a tower. He does not say which, but there no reason to doubt that it was the famous Leaning Tower.

In these experiments, Galileo showed that bodies did not fall at a rate that was proportional to their weight, as taught by Aristotle. Instead he believed, as Archimedes too had done, that the rate of fall of bodies might be proportional to their densities, since it seemed to him that though heavy balls started out more slowly than light balls, they overtook them during the descent – which is exactly what he would have observed. (What he did not realize – and what kept the true relation from him at the time – was that what he observed was a physiological effect: when an experimenter holds two unequal weights, palms downward, with outstretched arms, it is not possible to release the two weights simultaneously. The hand holding the heavy weight invariably opens a short time after the one holding the light weight – apparently an effect of differential muscular fatigue.[6] Today, this can be easily demonstrated by stop-action photography.) We can sympathize with Galileo – as for that matter with Aristotle. The study of terrestrial motion was long frustrated by the fact that the eye is (to express the point somewhat anachronistically) a digital device with rather limited capacity. The ability to freeze-frame the descent of a vertically dropped ball is beyond its powers. Ironically, the motions of the planets – which, because of their remoteness, appear effectively as if in slow motion – were better understood than those of dropping balls or projectiles or for that matter the detailed action of a horse's trot. The modern study of terrestrial motion – and the discovery of the proper law of acceleration – began in Padua a few years later, when Galileo realized he could slow the rate of motion of falling bodies to the point where the eye could follow what was happening by rolling them down inclined planes rather than by dropping them vertically from towers, but the detailed action of a horse's trot was not revealed until the nineteenth century, with the introduction by Eadweard Muybridge of stop-action photos.

In 1592, Galileo escaped from the oppressions of Pisa and took up a position in the more liberal city of Padua, which enjoyed the protection of the fiercely independent Serene Republic of Venice. By now he had a rather larger family to support than one might expect of a bachelor (he was not only responsible, as head of the family after Vincenzio's death in 1591, for his younger sisters and a brother, he was steadily acquiring additional dependents – two daughters, born in 1600 and 1601,

[6] As noted by Thomas B. Settle, "Galileo and Early Experimentation," in: Rutherford Aris, H. Ted Davis, and Roger H. Stuewer, eds., *Springs of Scientific Creativity*. Minneapolis: University of Minnesota Press, 1983, 3–20.

and a son, born in 1606, by his Venetian mistress, Marina Gamba). He was a passionate, earthy man, and hardly an ascetic; all his life he savored good food and fine wine (and with great regret had to give up those pleasures in advanced age because of failing health). Nevertheless, though he was often sailing close to the wind financially and had to supplement his salary as chair of mathematics by boarding students and selling copies of the manual for his military compass, he looked back on the 18 years he spent in Padua as the happiest period of his life. As he later confided to the Medici secretary, Belisario Vinta:

> Here I have a salary of one thousand florins for life, and this is perfectly secure, coming to me from a deathless and immutable ruler. I can earn more from private instruction as long as I care to go on teaching gentlemen from abroad, and if I were so inclined I could lay aside this much and more every year by taking scholars into my house. Moreover, my obligations here do not detain me more than sixty half-hours a year, and even then not so strictly that I cannot get in many free days; the balance of my time is perfectly free, and I am absolutely my own master.[7]

His personal happiness and the liberality of his position were reflected in his immense creativity and productivity during this period. Of course, his most important work was his elucidation of the laws of motion by rolling balls down inclined planes (as noted above). He did not publish it, however, until the last decade of life. In part, this was because he was diverted by the unforeseen opportunity of the telescope. With that, and fatefully, his Paduan idyll came to an end.

At almost the very moment when Galileo, the Doge, and the Senators were gathered on a tower in Venice to look through Galileo's telescope, Thomas Harriot, in England, was pointing a Dutch-made "perspective tube," magnifying 6×, at the Moon. So far as is known, Harriot's was the first astronomical use of the telescope.

Harriot, at 50, was 5 years older than Galileo, and came to the Moon as a seasoned explorer of another New World. A native of Oxfordshire, he was clearly a man of parts – mathematician, navigator, surveyor, cartographer, linguist. In 1583, he had joined Walter Raleigh's household in London. There he might well have met Giordano Bruno, just moved to London as a guest of the French ambassador and, after his visit to Oxford, returning to London in 1584 to pen his greatest work, the "Ash Wednesday Supper," affirming not only the reality of the heliocentric theory but suggesting that the universe was infinite and constituted of innumerable worlds.

In 1585, Bruno returned to Paris, while Harriot set sail as Raleigh's agent to establish a colony in Virginia. Had it survived, it would have been England's first foothold on the American continent, preceding Jamestown by more than a decade.

[7]Letter from Galileo to Belisario Vinta, 1610; quoted in Stillman Drake, *Discoveries and Opinions*, pp. 62–63. Drake notices that the sixty half-hours would seem to be in error. As the University records indicate that Galileo lectured daily at 3 o'clock in the afternoon during most of his years at Padua (*Opere*, xix, 119–120), Drake suggests that either the word "year" should have been "month," or the word "sixty" should have been "six hundred."

Harriot skillfully described the adventure in *A Briefe and True report of the New Found Land of Virginia*, a work that provides detailed accounts of the flora and fauna he encountered there, including maize and sunflowers and tobacco. Harriot also described the natives with the kind of detached and dispassionate observation he might have applied to the inhabitants of the Moon. He innovatively employed shorthand symbols to note the sounds of their speech in the hope of eventually writing a dictionary of tribal languages. Though he never did, his habit of thinking in terms of symbols perhaps led him to attempt the same with the French mathematician Vieta's seminal (but prolix) book on algebra. In any case, he introduced symbols for everything in the equations except actual numbers, a practice wherein mathematicians have, of course, followed him ever since.

After his return from America, Harriot participated, with Raleigh and the brilliant iconoclastic poet Christopher Marlowe, in intellectually adventurous meetings that Shakespeare caricatured in *Love's Labour's Lost* as the "School of Night." Among other things, they were accused of enjoying the dangerous pleasures of tobacco, of scoffing at the Bible, and of spelling the name of God backward. Marlowe – whose tempestuous career ended violently in 1593, in a tavern at Deptford; he was only 29 – regarded the miracles in the Bible as cheap juggler's tricks. Perhaps he was more impressed with Harriot's scientific experiments.

Harriot moved to the Abbey of Molanna near Youghal, County Waterford, and made a fortune managing Raleigh's vast Irish estates. He then entered the Earl of Northumberland's service, for which he received a handsome pension and a lifelong interest in Northumberland's land holdings at Brampton (county Durham) and in his estate of Syon House, east of London on the Thames near Kew; much modified in the eighteenth century, it still stands. Unfortunately, after the accession of James I, Harriot's patrons began to fall from favor. With Raleigh, whom James suspected of plotting to dethrone him, and Northumberland, whose kinsmen were involved in the Gunpowder Plot to blow up Parliament in 1604, Harriot was imprisoned in the tower of London. However, Harriot seems not to have been deemed very dangerous, and was soon released. He returned to the life of solitary and useful research he seems to have preferred. "I was … contented," he once wrote, "with a private life for the love of learning that I might study freely."

In pursuit of this private life of learning, he observed a comet (an apparition of Halley's) in 1607, the year Jamestown was founded by the Virginia Company, and may well have followed with interest reports of another Virginia Company expedition which set sail from Plymouth in early June 1609 but on July 24 was scattered by storm.[8] At almost that very moment Harriot first turned his "trunke" at the Moon, and made a crude sketch. Regrettably he entered no "Brief and True report of the New Found Land in the Moon" to parallel his earlier work on Virginia; indeed, his

[8]The Virginia expedition returned safely to Jamestown – all except the flagship, the *Sea Adventure*, which was presumed lost. In May 1610, two pinnaces appeared at Jamestown carrying the missing crew, who had run aground in Bermuda, which they found a delightful and hospitable place (belying its repute as an "Isle of Devils"), and slowly made their way back to rejoin their compatriots. Accounts of the *Sea Adventure* created a sensation and provide background to Shakespeare's *The Tempest*, which might better be described as *The Island*.

lunar observations were never published in his lifetime. With two friends in The Tower of London, he presumably wanted to keep a low profile. "The implacable stars are silent," he once said. Like himself it would seem.

His lunar observation was recorded at 9 p.m. on the evening of July 26, 1609 (Old Style; New Style August 5). Even allowing for the fact that Harriot's instrument cannot have been of very good optical quality, his drawing, made with short ragged strokes, hardly merits even the name of a sketch. L. Scott Montgomery states bluntly: "The talent was lacking."[9] For Harriot, the Moon was an ambiguous stimulus that was still regarded in Aristotelian or Dantean terms as a translucent alabaster ball or "vaporous" cloud. Harriot, in short, did not know *how to see* the Moon. He lacked the perceptual framework to decipher the image floating ambiguously before his eye and the artistic skill to set it down on paper. Moreover, Harriot's eye was literary not artistic. In Elizabethan England, as art historian Samuel Y. Edgerton, Jr., points out, literature flourished – it was, after all, the age of Marlowe and Shakespeare – but "the visual arts still languished in a sort of retardataire Gothic survival." Though Italian villains (inspired by the Florentine Machiavelli's *Prince*) were regularly being trotted out at the Globe in London in Shakespeare's plays, Italian influences in art had hardly yet been felt in England.[10]

All this is by way of explaining – and it does indeed demand explanation – the striking difference between Harriot's feeble effort and Galileo's superlative sepia ink wash drawings of the Moon produced only four lunations later. The first of

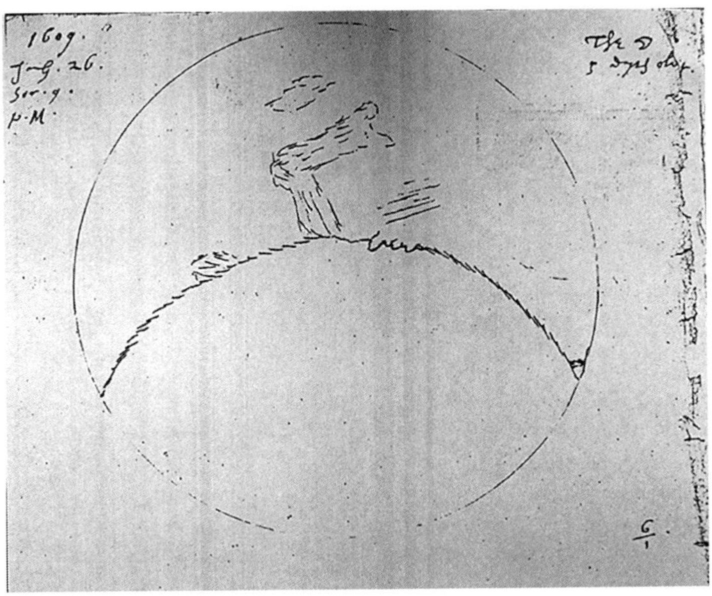

Thomas Harriot's drawing of the Moon

[9]L. Scott Montgomery, *The Moon and the Western Imagination.* Tucson: University of Arizona Press, 1999, p. 110.

[10]Edgerton, "Galileo, Florentine 'Designo'...", p. 228.

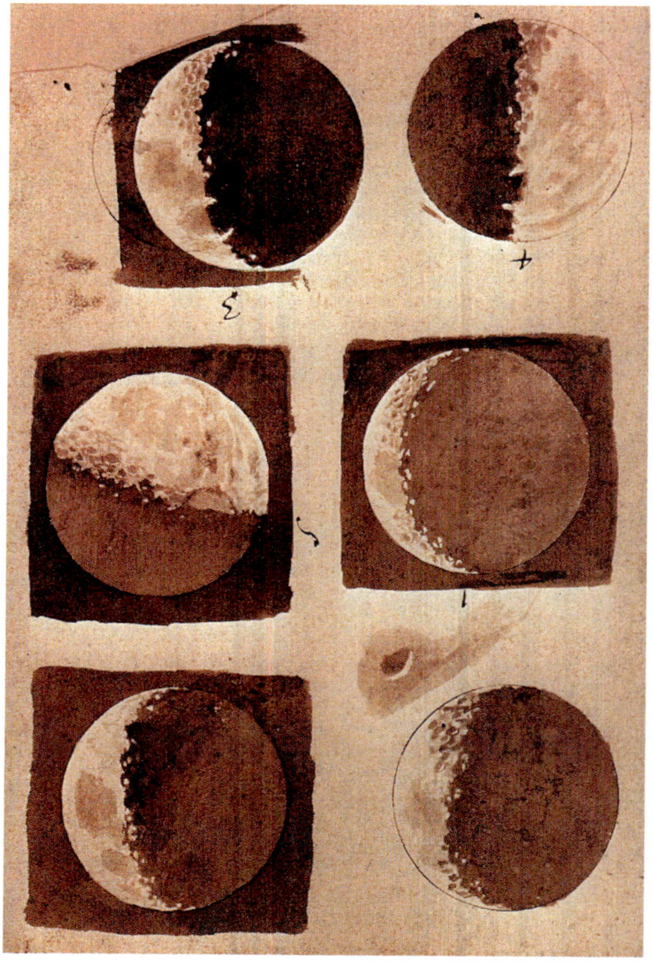

Ink wash drawings of the Moon by Galileo. Bibliotecha Nazionale Centrale, Florence

these, showing the 4-day Moon, has been dated with confidence by Ewen A. Whitaker to November 30, 1609.[11] With his remarkable manual dexterity and aptitude for tinkering, Galileo, since the previous summer, had continued to introduce improvements to his telescope – sparing "neither labor nor expense." He had painstakingly sifted through a vast quantity of lenses to find one that was optically superior to the rest, and had further improved the definition by stopping down the aperture so the light would pass only through the optically most perfect central part

[11]Ewen A. Whitaker, "Galileo's Lunar Observations and the Dating of the Composition of *Sidereus Nuncius*." *Journal of the History of Astronomy* 9 (1978), 155–169.

of that optically superior lens.[12] In this way he managed to produce an instrument magnifying 20×. Instead of something akin to an opera-glass or spyglass, he now had a real telescope – an instrument that could be used for pure research. Though still hampered by an annoying amount of chromatic aberration, as all refracting telescopes in the sixteenth and seventeenth centuries would continue to be, and inconvenienced by an exceedingly narrow field of view, he had produced for himself by far the most powerful optical appliance in existence at the time.[13]

Understandably, Galileo was fearful of potential competition, and eager to maintain a monopoly on telescopic observations for himself. He jealously guarded the secrets of the design, fabrication, and use of his instruments.[14] Though he had clearly won what telescope historian Albert van Helden has described as "the instrument race,"[15] the superior power of his telescopes was not the only factor accounting for his ability "to monopolize telescopic astronomy for the first several years and make almost all the important discoveries."[16]

The other factor was Galileo himself. Remember that in the seventeenth century the only detector available was the human eye, and Galileo's been superbly trained. His was an artist's eye, and he also had an artist's hand. A mere handful of successors in the next four centuries have rivaled him in eye-hand-coordination and artistic skill. He was not only able to see but, thanks to his early training, to imitate nature, to copy it as patiently and faithfully as possible with the techniques available to him.

Indeed, it appears that, after the early period in which he had thrown himself into drawing, Galileo had retained – as his biographer Viviani would say – "such a natural

[12] Stillman Drake, in *Galileo at Work: His Scientific Biography* (Chicago: University of Chicago Press, 1978), p. 148, suggests that the idea of stopping down the aperture may have occurred to Galileo because he had learned to peer through his clenched fist or between his fingers because of an eye condition. As we have seen, as early as Galileo's teen years, his father had removed him from the Benedictine monastery at Vallombrosa because of an inflammation of his eyes, and in Padua, while seeking to escape the heat during a visit to the countryside with friends, he contracted a severe illness from which one died immediately, another died later, and only Galileo survived. Dava Sobel, in *Galileo's Daughter*, p. 22, describes the immediate symptoms as "cramps and chills, intense headache, hearing loss, and muscle lethargy," while chronic symptoms suffered by Galileo included bouts of pain later described as "arthritic seizures." But he also suffered from problems with his eyes, which sound like early stages of glaucoma: "As a result of a certain affliction," he says (quoted in Sobel, p. 85), "I began to see a luminous halo more than two feet in diameter around the flame of a candle, capable of concealing from me all objects which lay behind it. As my malady diminished, so did the size and density of this halo, though more of it has remained with me than is seen by perfect eyes." Of course, in old age he went completely blind. The often-repeated claim that his blindness was due to his observing the Sun without protection seems to be without basis.

[13] See *Galileo's Telescope: the instrument that changed the world*, edited by Giorgio Strano (Florence, Istituto e Museo di Storia della Scienza, 2008).

[14] For a full discussion of this, see: Mario Biagioli, *Galileo's Instruments of Credit: telescopes, images, secrecy*. Chicago: University of Chicago Press, 2006.

[15] Albert van Helden, "Galileo and the Telescope," in: Paolo Galuzzi (ed.), *Novità celesti e crisi del sapere* (Florence: Giunti, 1984), p. 155.

[16] Mary Winkler and Albert van Helden, "Representing the Heavens: Galileo and Visual Astronomy," *Isis* 83 (1992), 214–216.

and proper inclination to the art of drawing, and in time acquired such exquisite taste, that his opinion on paintings and drawings was preferred to that of the foremost professors – even by the latter themselves."[17]

"Exhibit A" is the series of ink wash drawings he made of the Moon. On one page, Galileo has made six of them. The first four (according to the dates established by Whitaker) show the Moon between November 30 and December 2, 1609; two others represent the Moon on December 17–18. On another page is a seventh drawing, on which he also has calculated a horoscope. According to a close analysis by Horst Bredekamp:

> Galileo's drawings are remarkable not only for their precision, but also for their technique, using a brush to render the plasticity of the Moon's surface. All of the circles, with diameters of 57–59 mm, are drawn with pen compasses; in each case, the center point is marked by a tiny brown dot.... The use of brown color applied in differing densities made it possible to modulate from a deep, shadowy tone to a beige that almost fades to white.[18]

We cannot know for sure, but I would like to believe they were produced *en plein air*, and so record the magic of his first impressions. That would perhaps account for the peculiar orientation of the ink washes on the sheet. The upper middle one, showing the 4-day old Moon, was the first completed, then Galileo made the one to its right, but afterwards he turned the paper around and continued his series with the one below it. That artless arrangement seems unlikely if he were working them up in the studio from a series of roughs.

In any case, Galileo's drawings of the Moon (not in their original form but as engravings) were to produce a complete reorientation of mankind's views of the lunar surface. Henceforth the Moon no longer belonged to the distant and inaccessible heavens. Instead of the pure and perfect alabaster sphere of Aristotle and Dante, it became another earth. Though Galileo had seen no previous naturalistic drawings of the Moon (and hardly any exist), he was still able to approach it with eyes that had been trained by his great predecessors in Florentine art. Giotto had taught him how to use light and shade to suggest depth and solidity of the figures in his frescoes (as in the Scrovegni chapel in Padua), giving "the illusion of being able to touch them";[19] Masaccio, in the frescoes in the Brancacci Chapel of the Church of Santa Maria del Carmine in Florence, had given "tactile values to retinal impressions,"[20] and so on. He would of course have studied these frescoes carefully, and knew the way these techniques – these artist's conjurer's tricks – worked. Consider, for instance, Masaccio's fresco showing St. Peter healing the sick with his shadow: in sharp contrast to Medieval art, which is notoriously flat and in which the figures often cast no shadows, Masaccio has made the shadow something physical – it touches the sick, and is an active force. Similarly, the shadows of Galileo's ink wash drawings magically tease from the flat page the solidity of the lunar surface.

---

[17] Viviani, "Racconto," P. 602.

[18] Bredekamp, "Gazing Hands and Blind Spots," p. 172.

[19] Bernard Berenson, *Italian Painters of the Renaissance*, p. 40.

[20] Ibid.

Masaccio: St. Peter healing the sick with his shadow. Photograph by William Sheehan, 2008

Galileo later explained something of his method in a letter to his dear old friend Cigoli in 1612: "Of the objects appearing and seen, we see nothing but their surfaces; their depth cannot be perceived by the eye because our vision does not penetrate opaque bodies…. We know of depth, not as a visual experience per se and absolutely but only by accident and in relation to light and darkness."[21] It is a commonplace that through Galileo astronomy became a visual science. But as artists such as Giotto and Masaccio created the illusion one could reach out and touch their figures, Galileo not only showed the Moon and the objects of the heavens but made them almost tangible (as we will see, he did that also with Venus, in recognizing its phases as patterns of light and shadow on a sphere). One consequence of this is that he used simple trigonometry to deduce from the lengths of the shadows the heights of the lunar mountains. All that is quite an achievement for Galileo's art.

[21] Quoted in Edgerton, *The Heritage of Giotto's Geometry*, p. 225.

The Moon at the feet of the Virgin, painted by Galileo's friend Ludovico Cigoli between 1610 and 1612 on the ceiling of the Pauline Chapel in the Basilica of Santa Maria Maggiore, Rome. Photograph by William Sheehan, 2008

The haunting beauty of these ink washes was – as often proved the case with astronomical art over the centuries – lost in the engravings of them, even though Galileo himself supervised their creation. But Galileo did not trust to images alone to convey his eyepiece revelations; he also set out to capture in words what has been described as "the celestial theater of light and shadow opening up before his eyes"[22]:

> On the fourth or fifth day after new moon, when the moon is seen with brilliant horns, the boundary which divides the dark part from the light does not extend uniformly in an oval line as would happen on a perfectly spherical solid, but traces out an uneven, rough, and very wavy line as shown in the figure below. Indeed, many luminous excrescences extend beyond the boundary into the darker portion, while on the other hand some dark patches invade the illuminated part. Moreover a great quantity of small blackish spots, entirely separated from the dark region, are scattered almost all over the area illuminated by the sun with the exception only of that part which is occupied by the large and ancient spots. Let us note, however, that the said small spots always agree in having their blackened parts

[22] Bredekamp, "Gazing Hands and Blind Spots," p. 176.

> directed toward the sun, while on the side opposite the sun they are crowned with bright contours, like shining summits. There is a similar sight on earth about sunrise, when we behold the valleys not yet flooded with light though the mountains surrounding them are already ablaze with glowing splendor on the side opposite the sun…[23]

Henceforth Galileo's set of conventions for "seeing" the Moon become so determinative that it would be impossible to believe that the lights and shades along the terminator of the Moon were merely an illusion and elaborate *trompe l'oeil*, as Father Christopher Clavius and other Jesuit astronomers seem to have thought – or *hoped* – for a time. Galileo had done for celestial space what the practitioners of perspective had done for the terrestrial space. In painting, according to Gombrich: "Whether we want it or not, the illusion [of depth] is there." And he adds: "It is important to be quite clear at this point wherein the illusion consists. It consists, I believe, in the conviction that there is only one way of interpreting the visual pattern in front of us. We are blind to the other possible configurations because we literally 'cannot imagine' these unlikely objects."[24] Galileo had the right visual framework to "see" correctly – Edgerton, quoting Gombrich, says "he brought to his telescope a special 'beholder's share,' an eyesight educated to 'see' the unsmooth sphere of the moon illuminated by the sun's raking light."[25] Henceforth, every amateur with a telescope following the shadows of the lunar mountains crawling along the terminator (as I did as a 9-year old with my small telescope) or astronaut looking down on the Moon from a height of 70 nautical miles above the lunar surface has brought to the Moon the same special "beholder's share." We are all indebted to Galileo's insight. The Moon is the Moon he taught us to see.

[23]Galileo, *The Starry Messenger*; in Drake, *Discoveries and Opinions*, p. 32.

[24]Gombrich, *Art and Illusion*, p. 249.

[25]Edgerton, "Galileo, Florentine 'Disegno,' and the 'Strange Spottednesse' of the Moon," p. 227.

Chapter 9
Figures of Cynthia

Hither, as to their Fountain, other Stars
Repairing, in their gold'n Urns draw Light,
And hence the Morning Planet gilds her horns;
By tincture or reflection they augment
Their small peculiar, though from human sight
So far remote, with diminution seen.

Milton, Paradise Lost, VII, 364–369.

Galileo completed his first series of lunar observations – those recorded in the ink wash drawings – on December 18. By then he had also worked out the correct explanation of the phenomenon of earthshine – the dull-red glowing of the dark side of the young Moon.[1] He saw at once that earthshine was an illumination of the night-side of the Moon by sunlight reflected from the Earth. "The earth," he wrote, "in fair and grateful exchange, pays back to the moon an illumination similar to that which it receives from her throughout nearly all the darkest gloom of night." Having previously shown that the Moon – with its mountains and valleys – was another Earth, he could now demonstrate that the Earth was another planet, reflecting the sunlight from its continents and oceans onto the surface of the Moon.

By the end of his *annus mirabilis*, 1609, he had mapped multitudes of new stars in the Pleiades and Orion and discovered the true nature of the Milky Way:

> With the aid of the telescope this has been scrutinized so directly and with such ocular certainty that all the disputes which have vexed philosophers through so many ages have been resolved, and we are at last freed from wordy debates about it. The galaxy is, in fact, nothing but a congeries of innumerable stars grouped together in clusters.[2]

By the first week of January 1610, Galileo had fabricated a new, more powerful perspicillium. On January 7, he turned it to Jupiter and perceived (as he had not

[1] He was not, however, the first to do so. Leonardo had anticipated him, though his notes on the subject remained unknown. Kepler's teacher, Michael Mästlin, had discussed the same idea in a work on eclipses (1596), while Kepler had also made this identification in his treatise on optics (1604).

[2] Galileo, *The Starry Messenger*, p. 49.

W. Sheehan, *A Passion for the Planets: Envisioning Other Worlds, From the Pleistocene to the Age of the Telescope*, DOI 10.1007/978-1-4419-5971-3_9,

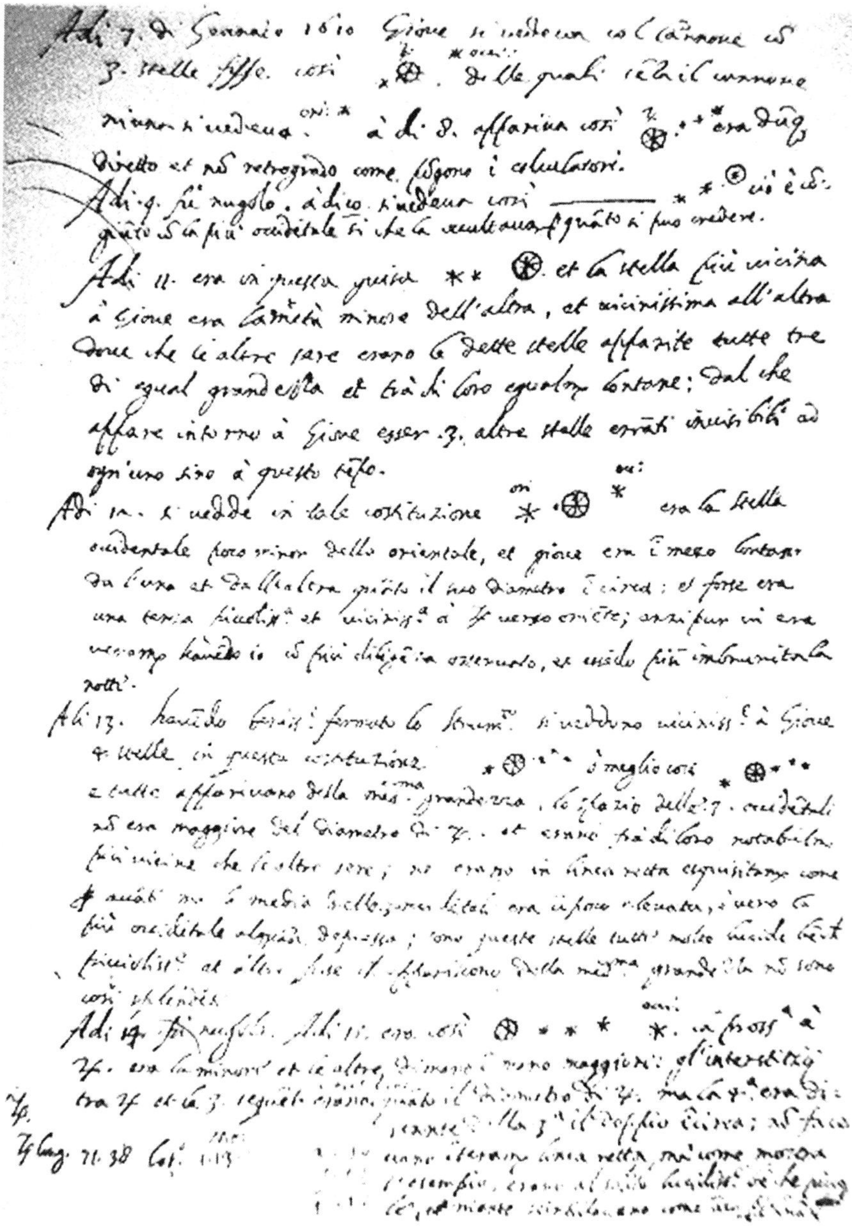

Sketches of Jupiter's satellites by Galileo. Bibliotecha Nazionale Centrale, Florence

with his "weaker" instrument) three starlets, small but very bright. They awakened his curiosity because – though at first he assumed they belonged among the fixed stars – they lined up perfectly along the ecliptic; two were on the eastern side and one on the western side of the planet, a lovely array. The next night, when he returned to the planet ("led by what, I do not know," he says disingenuously; of course, it was his insatiable curiosity) he found all three on the same (west) side of the planet and wondered whether – "contrary to the computations of the astronomers" – Jupiter was moving eastward at that time and so had got in front of them. He awaited the next night with anticipation. It was cloudy, alas. Finally on the tenth of January, he found only two of the small stars, to the east (the third, he assumed, was hidden behind Jupiter). He now realized that the phenomena he had witnessed could be explained only by assuming that the three small stars were actually small planets moving around the planet. He continued to follow them on succeeding nights. Finally, on January 13, he saw all four for the first time.

Galileo proceeded to write up his results at white-heat in the booklet *Sidereus Nuncius* (the Starry Message or, as usually translated, Messenger). Published in Venice on March 13, 1610 (the same day on which, 271 years later William Herschel would discover the planet Uranus), the first edition of 550 sold out almost immediately. L. Scott Montgomery points out that in naming himself a "messenger from the stars," Galileo is "being coy, humbly claiming to be only the receiver and deliver of this announcement. Yet he is also very much its author and sender, one who has returned from a distant place, ready to 'unfold great and very wonderful sights … to the gaze of everyone.'"[3] The engravings of the Moon (however far they fall short of the sepia ink wash drawings) and his vivid descriptions of the advancing sunrise over the lunar mountains and valleys furnished a vivid eyewitness account – nothing less than the impressionistic travelogue of one who has arrived in the New World in the Moon and written his report just as others (including Thomas Harriot) had issued reports of findings from distant journeys on Earth. Galileo's passionately argued report immediately captured the imaginations of ambassadors and poets. Indeed, on the very day *Sidereus Nuncius* made its appearance, the British Ambassador to Venice, Sir Henry Wotton, having got hold of a copy, wrote enthusiastically to King James I of England:

> Now touching the occurrents of the present, I send herewith unto His Majesty the strangest piece of news (as I may justly call it) that he hath ever yet received from any part of the world; which is the annexed book (come abroad this very day) of the Mathematical Professor at Padua, who by the help of an instrument (which both enlargeth and approximateth the object) invented first in Flanders, and bettered by himself, hath discovered four new planets rolling about the sphere of Jupiter, besides many other unknown fixed stars; likewise, the true cause of the *Via Lactea*, so long searched; and lastly, that the moon is not spherical, but endued with many prominences, and, which is all the strangest, illuminated with the solar light by reflection from the body of the earth, as he seemeth to say… The author runneth a fortune to be either exceeding famous or exceeding ridiculous.[4]

Though the "world in the Moon" had the most immediate effect on the imaginations of his contemporaries, Galileo himself regarded the new planets around Jupiter as his most important discovery. He now summoned up all his skill as a courtier to

[3] Montgomery, *The Moon and the Western Imagination*, pp. 116–117.

[4] Quoted in I. Bernard Cohen, *The Birth of the New Physics*, pp. 75–76.

exploit the opportunity by currying favor with the Medici family. Ultimately, he hoped to arrange his triumphal return to his beloved Florence, for as comfortable as was his position in Padua, he had long dreamed of returning to the Tuscan Court. As early as 1605, he had given private lessons on the use of his military compass, among other topics, to the Grand Duke, Cosimo II de'Medici . The previous summer, when Galileo demonstrated the use of one of his telescopes in Venice, Cosimo, on August 29, 1609, had written to express his interest in the new instrument and his desire to get one. Eventually Galileo presented him with a telescope – but not right away.

Only after he had discovered the New World in the Moon, the true nature of the Milky Way, and the satellites of Jupiter did Galileo, on January 30, 1610, contact Belisario Vinta, the Medici secretary, informing him that he had made a number of important discoveries (including the four "new planets") and asked permission to dedicate his discoveries to the Grand Duke. He proposed to name the satellites either the Cosimean or the Medicean Stars, wilily leaving the choice to his potential patron, and in the end, Cosimo preferred Medicean Stars, in order to apportion the honor between himself and his three brothers. (The names by which they are known today, Io, Europa, Ganymede, and Callisto, were proposed by a German astronomer, Simon Mayr, who claimed in a book published in 1614 that he had discovered them independently of – and a month before – Galileo. Inevitably, this led to a fierce priority dispute beginning with Galileo's accusation of plagiarism. The case continued to be debated for centuries, with adherents on both sides. The most likely interpretation is that Mayr saw the satellites independently but did not realize what they were until he read Galileo's pamphlet.)

When he published *Sidereus Nuncius*, Galileo trotted it out with a fulsome dedication to Cosimo. Galileo had spared no pains in working it up. On the back of the same sheet on which he had recorded his first set of six ink wash drawings of the Moon, Galileo had made a seventh lunar drawing showing the star Theta Librae emerging to the right (dated by Ewan Whitaker to January 19, 1610). In order to save paper, he also sketched out Cosimo's horoscope, showing Jupiter – the patron star of the Medici family – in reassuringly dominant position at Cosimo's natal hour. When he came to write his preface to *Sidereus Nuncius*, he laid it on thick by telling Cosimo that

> the Maker of the stars himself has seemed by clear indications to direct that I assign to these new planets Your Highness's famous name in preference to all others. For just as these stars, like children worthy of their sire, never leave the side of Jupiter by any appreciable distance, so (as indeed who does not know?) clemency, kindness of heart, gentleness of manner, splendor of royal blood, nobility in public affairs, and excellency of authority and rule have all fixed their abode and habitation in Your Highness. And who, I ask once more, does not know that all these virtues emanate from the benign star of Jupiter, next after God as the source of all things good? Jupiter; Jupiter, I say, at the instant of Your Highness's birth, having already emerged from the turbid mists of the horizon and occupied the midst of the heavens, illuminating the eastern sky from his own royal house [Sagittarius]….[5]

Though he had been treated well in Padua, as he confided to Belisario Vinta, Galileo had always longed for complete freedom from teaching and other formal duties in order to devote all his time to research. The closest model to what he had

[5] Galileo, *The Starry Messenger*, p. 25.

in mind was the position of court artist, someone who enjoyed "an independent position free of public teaching responsibilities and without specific work obligations."[6] In the end, Galileo was successful: he parlayed the spate of dramatic telescopic discoveries made at the end of 1609 and the beginning of 1610 into "his dream of a one-man institute for advanced study."[7] Within 3 months of the appearance of *Sidereus Nuncius*, Galileo was duly appointed Court Philosopher to the Grand Duke.

Naturally, the authorities in Venice were disappointed; but Galileo had crossed a Rubicon, and there was no turning back. With the benefit of hindsight, one can suggest – as his Venetian friend Giovan Francesco Sagredo would soon do[8] – that Galileo ought never to have risked leaving the safety of the Serene Republic, and that if he had remained in Padua he would never have been turned over to the Roman Inquisition. We can say that with the benefit of hindsight, but at the moment of his triumph, he must have felt as invincible as Michelangelo's Victory in the Palazzo Vecchio resting secure in the strength of his own right arm. His position at the Tuscan court would make him the "Michelangelo of mathematicians." Indeed, his salary – of 1,000 florins a year – was among the ten highest in Tuscany, and half again as great as that which had been paid to the prolific court sculptor Giambologna (died 1608; among his greatest works is the "Rape of the Sabines," in the Loggia dei Lanzi in Florence).

Galileo was now not only court mathematician but court philosopher and even court artist for he, and he alone, determined that in the future astronomy was to be first and foremost a visual science. Well did the English poet John Milton – who got so much else about Galileo wrong – characterize him as "the Tuscan artist."[9]

In preparing for the move to Florence, Galileo ended his relationship with Marina Gamba, who remained in Venice and before long entered a respectable marriage with one Giovanni Bartoluzzi. In July, just before his departure from Padua, Galileo made telescopic observations of Saturn which showed its triple-form (announced in an anagram which poor Kepler thought referred to the discovery of two satellites of Mars!).

Galileo arrived in Florence in September 1610, and while seeking a house "with a high terraced roof, from which the whole sky is visible," moved in temporarily

---

[6] Bredekamp, "Blazing Hands and Blind Spots," p. 156.

[7] Ibid., p. 170.

[8] *Opere* xi, 170–172. Translated by Drake, *Discoveries and Opinions*, pp. 66–68.

[9] The passage is from *Paradise Lost*, V, 287–291:

> … the moon, whose orb
> Through optic glass the Tuscan artist views
> At evening from the top of Fiesole,
> Or in Valdarno to descry new lands,
> Rivers or mountains in her spotty globe.

There is no record of Galileo's having observed the Moon from the lofty heights of the ancient city of Fiesole, which is located north of Florence. Florence itself is, of course, situated in Valdarno – the Valley of the River of the Arno – but Galileo's discoveries about the Moon were made in Padua, not Florence.

with his sister Virginia and her husband. Almost at once his health began to fail. He was ill throughout the winter – miserably complaining that "the very thin air of Florence" was "a cruel enemy of my head," and suffering from colds, discharges of blood, and constipation which reduced him "to such a state of weakness, depression, and despondency that I have been practically confined to the house, or rather to my bed, but without the blessing of sleep or rest."[10] Under these conditions he commenced the first series of telescopic observations ever made of the planet Venus, which had emerged into the evening sky in early August and over the next several months underwent before his wondering eyes marked changes in size and form from which Galileo would wrest a decisive triumph of the Copernican over the Ptolemaic theory.

According to Ptolemy, Venus always lies between the Sun and the Earth, and can never appear more than half-illuminated. According to Copernicus, it circles around the Sun, and can sometimes show as a gibbous or even as a fully illuminated disk (when it is on the far side of the Sun from the Earth).

By the time Galileo arrived in Florence, Venus was just becoming well placed for observation in the western sky, and Galileo must have examined it as soon as he unpacked his telescopes and settled his head. At the time, the planet was still far from the Earth; it would have presented as a dazzling but small disk, that might seem to be shining by its own light. Only after the middle of November – and especially in early December – did Venus become an exciting object. Both Galileo and his gifted pupil, Benedetto Castelli, monk of Montecassino, realized at nearly the same moment that if Venus could be shown to have phases like the Moon's, the truth of the heliocentric system might be dramatically demonstrated. On December 11 Galileo submitted an anagram – a scramble of letters announcing his suspicions about the planet – to Cosimo's brother, Giuliano de'Medici: *Haec immature a me iam frustra leguntur o y*. (The o and y at the end could not be used in forming the anagram.) In translation this says, "These premature from me are at present deceptively gathered together." Galileo revealed the solution to Giuliano only on New Year's Day 1611: *Cynthiae figures aemulatur mater amorum*, i.e., "The mother of loves imitates the figures of Cynthia [the Moon]." In other words, Venus passes through the whole gamut of moonlike phases in accordance with the Copernican but not the Ptolemaic system. With this discovery, Galileo had dealt a decisive death-blow to the Ptolemaic system.

Though Galileo always feared being upstaged by potential competitors, he had the telescopic examination of Venus to himself that autumn. Such was the superiority of his telescopes and of his scientific vision that there were as yet few observers who succeeded in even partly following his tracks. According to Giorgio de Santillana:

> Certain doctors, who at least had the courage of their convictions, did actually and steadfastly refuse to look through the telescope, as has been recounted many times. Some did

[10]Sobel, *Galileo's Daughter*, pp. 38–39.

> look and professed to see nothing; most of them, however, gave it the silent treatment or said they had never gotten around to looking through it but that they already knew that it would show nothing of philosophical value. One maintained that it was impossible that the ancients should not have had such instruments, since they had excelled in everything, and that their silence on the subject implied an unfavorable judgment of their performance.[11]

Though few went as far as Magini, professor of astronomy at Bologna, who vowed to "extirpate" the new planets from the sky, some untrained observers used such poor telescopes that they were beguiled or disappointed with mirages and deceptions (such as "quadratic" stars). Still others, without training in either seeing or drawing, were simply baffled, as was Thomas Harriot's associate, Sir William Lower, a Welshman, who confessed that when observing the "Seven Sisters" he did see eight but did not dare trust what he saw "because I was prejugd with that number, I beleved not myne eyes."[12]

Apart from the problematic case of Mayr, the first to certifiably see the "Medicean stars" was Cosimo himself, to whom Galileo revealed them with a good telescope at Pisa in April 1610. Next was Kepler who after initial failures (remember, his eyesight was very poor), finally glimpsed them on several nights in August with a telescope Galileo himself had fashioned for the Elector of Cologne. Then, in November, they were seen by Harriot in England, by Nicolas-Claude Fabri de Peiresc and Joseph Gaultier de la Vallette in Provence, and – most importantly – by the at-first skeptical Father Clavius and other Jesuit mathematicians at the Collegio Romano.[13] By then, of course, Galileo himself had moved on to new discoveries, and was making the observations described above of the phases of Venus.

By March 1611, Galileo had arrived at the height of his fame, and set out on a journey from Florence to Rome, stylishly carried in a litter of the Grand Duke. At each point along the way – at San Casciano, Siena, San Quirico, Acquapendente, Viterbo, Alonterosi – he set up his telescope and added to his series of observations of the Medicean Stars. But in Rome, where he demonstrated his telescope at the Collegio Romano, the Jesuits' own enclave, he was already coming under scrutiny of wary Church authorities. His enthusiasm, and perhaps his arrogance, blinded him to the dangers. The Florentine ambassador, Piero Gucciardini, a seasoned observer of power politics, saw Galileo as "all afire in his opinions, and full of great

[11]Santillana, *The Crime of Galileo*, pp. 7–8.

[12]Quoted in Terrie F. Bloom, "Borrowed Perceptions: Harriot's Maps of the Moon." *Journal for the History of Astronomy* 9 (1978):117–122:121.

[13]The superiority of Galileo's instruments is attested to by the case of Peiresc, who used a telescope by Jacob Metius, who had been one of the original claimants to invention of the telescope. Peiresc admitted to Galileo in 1634 that though Metius' instrument allowed him to just glimpse the four satellites of Jupiter, it was not powerful enough to allow him to carry out the study he had hoped to make with it, the precise determination of the satellites' orbital periods. He asked whether Galileo would be able to produce and send him a good telescope.

passion, but without strength or prudence to control himself."[14] He put it even more bluntly on another occasion: "This is no fit place to argue about the Moon."

Gucciardini was right. Between a passionate seeker for truth like Galileo and an authoritarian and dogmatic institution like the Church, there could be no compromise – it could only be, as Jacob Bronowski has said, that "for twenty years and more he [would move] along a path that led inevitably to his condemnation... The division between him and those in authority was absolute. They believed that faith should dominate; and Galileo believed that truth should persuade."[15] Galileo's advocacy of the heliocentric system was viewed with alarm by Cardinal (later Saint) Robert

Cardinal Bellarmine

[14]Santillana, *The Crime of Galileo*, p. 124.

[15]Jacob Bronowski, *The Ascent of Man* (Boston: Little, Brown & Co., 1973), p. 205.

Bellarmine, a Jesuit and the Vatican's chief theologian, who had signed the order condemning Bruno to the stake. Bellarmine was a Scriptural literalist. He believed that Scripture had to be interpreted (except as a last resort) in the literal sense. He was loath to make exceptions, and insisted even on passages such as "The Sun also riseth, and the Sun goeth down, and hasteth to his place whence he arose," which made the Sun not the Earth move. Both Bellarmine and his superior, Pope Paul V – no intellectual, he was by temperament doctrinaire and inflexible – were inclined to declare Copernicanism heresy and to place *de Revolutionibus* on the Index of Forbidden Books.

Meanwhile, at the end of 1611 and at the beginning of 1612, Galileo began observing sunspots with his telescopes. He disputed priority over their discovery and interpretation with Jesuit astronomers, including Christoph Scheiner of Ingolstadt, in the *Letters on Sunspots* (published in 1613). As usual, his rapier-like wit was devastatingly effective, and he managed to win their implacable enmity. Though their opposition to him was gathering behind the scenes, he seemed to have nothing to fear when, in February 1616, he was summoned to Bellarmine's palace in Rome. Bellarmine issued a stern warning: he was "not to hold or defend" the heliocentric theory. For the next 7 years, living in comfort in the palatial Villa dell'Ombrellino on the hill of Bellosguardo, on the south side of the Arno, he obeyed. As long as Paul V remained Pope, he must have seen that it was prudent to keep silent.

By 1623, everything seemed to have changed; Paul V had died – entombed, ironically, under the craterous Moon Galileo's old friend Cigoli had painted in the Pauline Chapel in Santa Maria Maggiore. After the brief interlude of an elderly

Collegio Romano. Photograph by William Sheehan, 2008

Bellosguardo, Galileo's residence in Florence. Drawing by Randall Rosenfeld. © Randall Rosenfeld

caretaker pope, Gregory V, Cardinal Maffeo Barberini, an intellectual and friend of Galileo, assumed the triple crown as Pope Urban VIII.

Barberini was a member of the Academy of Lynxes, the world's first scientific society devoted to encouraging observation and experiment. Founded by Federico Cesi, Marquis of Montecelli and son of the Duke of Asquaparta, Galileo was its

Urban VIII, as painted in about 1632 by Bernini. Nazionale d'Arte Antica, Rome. Photograph by William Sheehan, 2008

most illustrious member. In 1616, Barberini had worked behind the scenes to persuade Paul V and Bellarmine not to proscribe *de Revolutionibus* but to allow it to pass after a few minor revisions. The Church still took the official view that Copernicus had authored the preface suggesting that the heliocentric system was only a hypothesis. But Barberini's elevation to the papacy raised expectations among liberal Catholics such as Galileo. In fact, one of the pope's first priorities was to call Galileo to Rome. They discussed astronomy.

On the critical point Urban proved to be, alas, no closer to Galileo's position than Paul V or Bellarmine had been. He too subscribed to the view of Osiander: the heliocentric system was only a hypothesis, and it could never be more. It was not for man to dictate the limits of God's power. Even though Galileo's demonstrations were sound, it was still possible for God to have arranged things in an entirely different manner than grasped by the astronomers, and to bring about the effects seen in the heavens in a manner that preserved the literal truth of the Scriptures. Galileo failed to take the measure of his erewhile friend: the intellectual Cardinal became an authoritarian and doctrinaire pope. His absolutist dream was to wipe out Protestantism everywhere just as that of Magini had been to extirpate the Medicean stars from the sky. In the larger picture of his concerns, he did not have time to argue with an insubordinate scholar.

From the moment Urban became pope – if not, indeed, from his first visit to Rome in 1611 – Galileo's course, though intricate as that of the Evening Star passing through its phases from full to waning crescent, was fixed and irrevocable as fate. Mistakenly presuming on the literalism of Bellarmine's 1616 decree – that as long as he did not "hold and defend" the heliocentric system, he could still teach it – Galileo, at Bellosguardo, began to write the *Dialogue Concerning the Two Chief World Systems*.

After he had finished it, he moved from Bellosguardo to Il Gioello, the "Jewel," a villa in Arcetri, where he prepared it for the press. He had moved to Arcetri so that

Il Gioello. Drawing by Randall Rosenfeld. © Randall Rosenfeld

Refectory of the Convent of San Matteo, Arcetri, where Galileo's daughters were sequestered nuns. Most of the convent has been rebuilt following destruction by fire, and the refectory alone survives from Galileo's time. Photograph by William Sheehan, 2008

he could be closer to his daughters, of whom the eldest, Suor Maria Celeste, was especially dear to him. She was sequestered in the Convent of San Matteo just down the hill from Il Gioello, and their relationship – which Dava Sobel has spellbindingly told in *Galileo's Daughter* – was the chief prop and comfort of his later years.

Dutifully cleared by censors, the *Dialogue* was published in Venice in 1632. Taking the form of a dialogue among characters debating arguments for and against the Copernican and Ptolemaic systems – the Tychonic, the favorite refuge of the Jesuits since it also gives the correct interpretation of the phases of Venus, is not mentioned – the book is far from even-handed. Even so it might have passed muster had Galileo not imprudently given to Simplicio, an Aristotelian who really does come across as rather a simpleton, the very argument about man not putting limits to God's power that Urban himself had made in 1623. Hell hath no fury like a Pontiff scorned, and in his fury, Urban unleashed the dogs of the Inquisition.

The Gregorian Tower of the Winds, which stands above the Vatican's secret archive where the documents used to frame Galileo were kept. Photograph by William Sheehan, 2008

Though there were technical problems with the prosecution, there was never any question about the result. By resorting to the methods of the police state, the Inquisition surprised Galileo with secret evidence (as was only revealed in 1877 when the contents of the Vatican's secret archives were finally divulged). Without being able to examine the basis of the case against him, he was duly prosecuted – and found guilty – on "vehement suspicion of heresy."

On June 21, 1633, Galileo appeared in the Convent of the Church of Santa Maria Sopra Minerva (located on the square behind the Temple to the Planetary Gods or Pantheon built by Hadrian in Ptolemy's time). His head bowed and bent on arthritic knees, he was forced to "adjure, curse, and detest" his errors. He was sent to the Villa Medici, and for 2 months awaited his trial, which would determine the Inquisition's decision as to his fate. At this point he could still have been imprisoned or even tortured. In the end, the imprisonment was commuted, and he was allowed leave Rome. After spending several months under supervision of the Archbishop of Sienna, Ascania Piccolomini, a learned and sympathetic man to whom he demonstrated his telescope, he was granted leave to return to Tuscany – though, except for medical treatment, he was never again allowed within the city walls for the rest of his life. Under virtual house arrest in Arcetri, he produced his greatest book, the *Discourses*

Solar Tower, Arcetri. This observatory, built under the direction of Giovanni Battista Donati in the nineteenth century, is located only a short walk from Galileo's home of Il Gioello. Photograph by William Sheehan, 2008

on the New Sciences, which revealed the principles of the motion of bodies discovered long before in Padua, and in July 1637 – with the feeble and fading vision of his left eye; the right was already blind – he made his last discovery, the libration of the Moon. He died at Arcetri on January 8, 1642; by then he was completely blind.

Only now, in death, was he allowed to return to Florence. Persecuted still by the relentless Urban, his body was not allowed to rest in the nave of the Basilica of Santa Croce, where the then-Grand Duke of Tuscany, Ferdinand II, wanted to bury it, but in a small room next to the novices' chapel under the bell tower. Galileo's mortal remains were finally disinterred from that spot in 1737, and moved to an ornate tomb near the entrance and across the nave from Michelangelo. The condemned *Dialogue* as well as uncensored versions of *De Revolutionibus* by Copernicus would remain on the Catholic Church's Index of Forbidden books until 1835. Not

Tomb of Galileo, the Church of Santa Croce, Florence. Photograph by William Sheehan, 2008

until 1893 did the Catholic Church belatedly decide that it was no longer heresy to affirm the truth of the heliocentric system, while only in 1992 – 350 years after Galileo's death – did John Paul II – in a long-delayed gesture of magnanimity – finally pardon Galileo of his heresy.

As a pilgrim, I have followed Galileo to many of the places made legendary by his association with them. I have sought his presence in Rome, among the palaces of the Caesars and the temples of antiquity ransacked by the Renaissance and Baroque popes in order to build their own monuments and tombs. I have walked the streets of Rome to the Collegio Romano, where Galileo enjoyed his triumph of 1611 and where he returned in 1616 to appear before Bellarmine. I have peered through the windows of the Vatican Museums at the papal gardens and tried to imagine Galileo's fateful discussion with Urban VIII in 1623 (and enjoyed the serenades of birds which in Urban's time were killed because they interfered with his sleep). I have stood in the Piazza of Santa Maria sopra Minerva behind the Pantheon and reflected on the melancholy scene of Galileo's humiliation before the Inquisition.

Though it is possible to find traces of Galileo's life in Rome, it is elsewhere – in Padua, Venice, and Florence – that he has left the strongest earthly impressions. Especially in Florence. There I have chased him up the hill of Bellosguardo, a fitting place to entertain cosmic views: the scene dominated by the sight of Brunelleschi's dome and Giotto's tower and the hill of Fesole, so lovingly evoked by its onetime resident Elizabeth Barrett Browning, who lived here in the nineteenth century:

> From Tuscan Bellosguardo,
> Where Galileo stood at nights to take
> The vision of the stars, we have found it hard,
> Gazing upon the earth and heavens, to make
> A choice of beauty.

I have taken the route up to Arcetri from the ramparts of Fort Belvedere, into the countryside alongside the narrow and enchanting Via San Leonardo between the enclosed walls and groups of fine houses, past the Observatory of Arcetri, founded by Giovanni Battista Donati in the nineteenth century, to Galileo's villa – still a quiet place – nestled below the Torre del Gallo on the hills in the near distance. I have visited the Convent of San Matteo where his daughters lived (only the refectory remains from Suor Maria Celeste's time).

I have paid my respects to his baroque tomb in Santa Croce, which contains his bones as well as those of Suor Maria Celeste. That gaudy thing is not, however, his true monument. That is found above – in the mountains and craters of the Moon, in Jupiter with its ever-rushing satellites, in Venus whose phases he first saw – all wonders he first revealed to the world with telescopes some of which are preserved in the Museum of Science along with a relic of one of Galileo's index fingers, still pointing upward to the sky after all these years.

One evening in Florence, I wandered across the Arno and watched the sunset. I could not help thinking of the experience of the nineteenth century French astronomer Camille Flammarion, in whose hands Donati once placed one of Galileo's precious telescopes:

> After sunset I recaptured the spirit of the Florentine astronomer on one of the beautiful Italian terraces just as the stars were coming out. With feverish impatience I turned this marvelous tube toward the new worlds that he had discovered in the heavens. I recalled that he had shown these sights to those who were incredulous; he still shows them to us today from his grave.[16]

[16]Camille Flammarion, *La Planète Mars*. Paris: Gauthier-Villars, 1892, vol. 1, p. 5.

Chapter 10
Afterglow

Sad Hesper o'er the buried sun
And ready, thou, to die with him,
Thou watchest all things ever dim
And dimmer, and a glory done.

Alfred, Lord Tennyson, In Memoriam

In the autumn and winter of 2008–2009, I observed the latest – my 28th – apparition of Venus as an Evening Star since the first spring apparition I watched when I was a 9 year old.

Conjunction of Venus and the Moon, July 2, 1921, as seen over Camille Flammarion's observatory at Juvisy-sur-Orge. Painting by Julian Baum. © Julian Baum

W. Sheehan, *A Passion for the Planets: Envisioning Other Worlds, From the Pleistocene to the Age of the Telescope*, DOI 10.1007/978-1-4419-5971-3_10,

I began to realize how much the rhythms of this lovely planet's appearances and disappearances in the Evening sky, alternating with its appearances and disappearances in the Morning sky, have woven their threads through the fabric of my life. The elongation just past was unusually favorable for Northern observers. As if following the trajectory of a human life, the planet struggled slowly away from the Sun at the end of the previous summer; through the months of winter it rose and grew stronger – for awhile it seemed invincibly steady and bright – as it looked ready almost to challenge the zenith.

For weeks its motion was almost imperceptible, as it paired with Jupiter and the Moon in early December and with the Moon in January and again in February. I thought of the youth and middle years of a man's life, when one feels strong and impervious to the effects of age.

Then the rapid decline, exactly like that impetuous rapidity with which – in the last decades of a man's life – time seems eager to be rushing through the neck of the hourglass. It abandoned its leisurely pace and at the end of March galloped with giant strides back toward the Sun, fading as it did so. Instead of being strong and steady as it had been, it grew feeble and twinkling. The twinkling was remarkable the last week or so before inferior conjunction, for despite the fact that the planet's apparent disk had grown to almost a minute of arc, it was only a few percent illuminated, and the thin and fragile arc was readily disrupted by the flailing currents of air. The apparition was now getting long in the tooth; it had entered its dying days.

I saw it the very evening before it reached inferior conjunction. For years earlier, it had transited Sun. Now it passed as far above the Sun as its orbital tilt can take it – a full 8° north of the solar disk. In the telescope it appeared as a delicate, trembling circle of light – a thing of dream not tangible reality – its illuminated hemisphere seemed to extend, an attenuated will-o'-wisp, all the way around the planet. Then it was gone, headed above and beyond the Sun and soon to be reborn in the morning sky, continuing the perennial drama of death and resurrection that seemed so rich in meaning to the ancients.

Inevitably, during the 400th year of Galileo's telescope, I thought of the great man peering at Venus shining bravely in the sky above the domes and towers and red roofs of Florence. I tried to imagine the *eureka*! moment, during the first week of December 1610, when he first realized that the planet's behavior was as it must be if Copernicus were right, as it transformed before his eyes from stout squat awkward gibbous into lovely svelte slender crescent.

From the pure serene of Bellosguardo, a few years after Galileo first captured Venus's waning phase against the pellucid Florentine sky, he wrote in *The Assayer* that in order to become science philosophy must

> throw out blind respect for authority … [and] learn to be content with pursuing limited objectives, reaching out gradually into the infinity of unknown events and undiscovered laws of nature, without ever achieving complete and exact knowledge of anything at all.[1]

[1] Drake, *Discoveries and Opinions*, p. 224.

That means that science can never be certainty or fulfillment, only aspiration and quest. Passion exists for what we desire but can never wholly possess. As I followed Venus through another elongation, I felt my old passion for the planet as strongly as ever. I hope to do so until the very day that – like the planet – having passed my own Greatest Elongation East of the Sun, I slope from the evening gloaming into whatever Inferior Conjunction must in the end quench and extinguish our doubtful and flickering human kind of star.

Index

C

D

E

F